Research Priorities in Tropical Biology

Committee on Research Priorities in
Tropical Biology
Division of Biological Sciences
Assembly of Life Sciences
National Research Council

NATIONAL ACADEMY OF SCIENCES
WASHINGTON, D.C.   1980

NOTICE: The project that is the subject of this report was approved by the Governing Board of the National Research Council, whose members are drawn from the Councils of the National Academy of Sciences, the National Academy of Engineering, and the Institute of Medicine. The members of the committee responsible for the report were chosen for their special competencies and with regard for appropriate balance.

This report has been reviewed by a group other than the authors according to procedures approved by a Report Review Committee consisting of members of the National Academy of Sciences, the Academy of Engineering, and the Institute of Medicine.

This study was supported by the National Science Foundation, Contract NSF-C310, Task Order No. 374.

Library of Congress Cataloging in Publication Data

National Research Council. Committee on
    Research Priorities in Tropical Biology.
    Research priorities in tropical biology.

    Chapter 1 in English, French, Portuguese,
and Spanish.
    Bibliography: p.
    1. Biological research—Tropics.
2. Ecological research—Tropics.   3. Tropics.
I. Title.
QH84.5.N37 1980   574.909′3′072   80-15773
ISBN 0-309-03043-9

Available from

Office of Publications
National Academy of Sciences
2101 Constitution Avenue
Washington, D.C. 20418

Printed in the United States of America

Committee on Research Priorities in Tropical Biology

PETER H. RAVEN, Missouri Botanical Garden, St. Louis, Missouri (*Chairman*)
PETER S. ASHTON, Harvard University, Cambridge, Massachusetts
GERARDO BUDOWSKI, Centro Agronómico Tropical de Investigación y Enseñanza, Turrialba, Costa Rica
ARTURO GÓMEZ-POMPA, Instituto Nacional de Investigaciones sobre Recursos Bióticos, Xalapa, Mexico
DANIEL H. JANZEN, University of Pennsylvania, Philadelphia
WILLIAM M. LEWIS, JR., University of Colorado, Boulder
HAROLD MOONEY, Stanford University, Stanford, California
PAULO NOGUEIRA-NETO, Secretário Especial do Meio Ambiente, Ministerio do Interior, Brasília, Brazil
GORDON H. ORIANS, University of Washington, Seattle
HARALD SIOLI, Max Planck Institut für Limnologie, Plön, Federal Republic of Germany
HILGARD O'REILLY STERNBERG, University of California, Berkeley
JOHN J. TERBORGH, Princeton University, Princeton, New Jersey
FRANK H. WADSWORTH, Institute of Tropical Forestry, Rio Piedras, Puerto Rico
PAUL J. ZINKE, University of California, Berkeley

JAMES J. TALBOT, NAS, *Staff Officer*

PANELS

Asian Ecosystem Site

PETER S. ASHTON (*Chairman*)*
ANGEL ALCALA, Department of Zoology, Silliman University, Dumaguete City, Philippines
MICHAEL GALORE, Office of Forests, Lae, Papua New Guinea
C. V. S. GUNATILLEKE, Department of Botany, University of Sri Lanka, Peradeniya, Sri Lanka
KUSWATA KARTAWINATA, Curator, Herbarium Bogoriense, L.B.N., Jalan Juanda Raya, Bogor, Indonesia
ENGKIK SOEPADMO, University of Malaya, Lembah Pantai, Kuala Lumpur, Malaysia
TEM SMITINAND, Royal Forest Department, Bangkok, Thailand
SOMSUK SUKWONG, Kasetsart University, Bangkok, Thailand
L. G. M. TANTRA, Forest Research Institute, Bogor, Indonesia

(This panel met in Bangkok, Thailand, in September 1979.)

Ecosystems

GORDON H. ORIANS (*Chairman*)*
PETER S. ASHTON*
DANIEL H. JANZEN*
CARL JORDAN, Institute of Ecology, University of Georgia, Athens
THOMAS LOVEJOY, World Wildlife Fund—U.S., Washington, D.C.
HAROLD MOONEY*
PETER H. RAVEN*
FRANK H. WADSWORTH*
PAUL J. ZINKE*

(This panel met in St. Louis, Missouri, in January 1979.)

Human Ecology

HILGARD O'REILLY STERNBERG (*Chairman*)*
BRENT BERLIN, Department of Anthropology, University of California, Berkeley

*Member of the committee.

Panels

PIERRE DANSEREAU, Université du Québec à Montréal, Montréal, Canada
ARTURO GÓMEZ-POMPA*
ROY A. RAPPAPORT, Department of Anthropology, University of Michigan, Ann Arbor

(This panel met in Washington, D.C., in December 1978.)

Limnology

HARALD SIOLI (*Chairman*)*
CHRISTIAN JUNGE, Max Planck Institut für Chemie, Mainz, Federal Republic of Germany
WILLIAM M. LEWIS, JR.*
J. F. TALLING, Freshwater Biological Association, Windemere Laboratory, The Ferry House, Ambleside, Cumbria, England
JOHN B. THORNES, Department of Geography, The London School of Economics and Political Sciences, University of London, London, England
FRANZ H. WEIBEZAHN, Escuela de Biología, Facultad de Ciencias, Universidad Central de Venezuela, Caracas

(This panel met at Cumberland Lodge, Windsor Great Park, England, in January 1979.)

Plant Physiological Ecology

HAROLD MOONEY (*Chairman*)*
OLLE BJÖRKMAN, Department of Plant Biology, Carnegie Institution of Washington, Stanford, California
A. E. HALL, Department of Botany, University of California, Riverside
ERNESTO MEDINA, Centro de Ecología, Instituto Venezolano de Investigaciones Científicas, Caracas, Venezuela
F. B. TOMLINSON, Harvard Forestry School, Harvard University, Petersham, Massachusetts

(This panel met at Stanford University, Stanford, California, in January 1979.)

*Member of the committee.

Contents

PREFACE ... ix

1 SUMMARY AND RECOMMENDATIONS ... 1

   Biological Inventory, 2
   Tropical Ecosystem Studies, 3
   Studies of Tropical Aquatic Systems, 4
      Rivers, 4; Lakes, 5; Wetlands, 5; Precipitation, 5
   Monitoring Forest Conversion, 5
   Five-Year Schedule, 6

1 EXPOSÉ SUCCINCT ET RECOMMENDATIONS ... 7

   Inventaire Biologique, 8
   Etudes des Systèmes Ecologiques Tropicaux, 9
   Etudes des Systèmes Tropicaux Aquatiques, 11
      Fleuves, 11; Lacs, 11; Terres Marécageuses, 11; Précipitations, 12
   Contrôle de la Conversion Forestière, 12
   Calendrier Quinquennal, 12

1 SUMÁRIO E RECOMENDAÇÕES ... 14

   Inventário Biológico, 15
   Sistemas de Estudos Ecológicos Tropicais, 16
   Estudos de Sistemas Aquáticos Tropicais, 17
      Rios, 18; Lagos, 18; Terras Alagadiças, 18; Precipitação, 19

Contents

Control da Transformação de Florestas, 19
Programa de 5 Anos, 19

1 RESUMEN Y RECOMENDACIONES 20

Inventario Biológico, 21
Estudios de Ecosistemas Tropicales, 22
Estudios de los Sistemas Acuáticos Tropicales, 23
  Los Ríos, 23; Los Lagos, 24; Las Tierras Húmedas, 24;
  La Precipitación, 24
Vigilancia de la Conversión Forestal, 25
Plan Quinquenal, 25

2 THE PROBLEM 26

Conversion Rates of Tropical Moist Forest, 29
Land Use, Population Growth, and Food Production, 30
Preservation of the Forests, 33
Scientific Rationale for Learning More about the Tropics, 35
Societal Rationale for Learning More about theTropics, 40

3 INVENTORY OF TROPICAL ORGANISMS 46

Number of Systematists, 47
Methods and Criteria for Sampling, 49
Evolutionary Systematics, 53
Historical Studies, 55
Preservation of Genetic Resources, 56
Critical Areas, 57

4 STUDIES OF SELECTED TROPICAL ECOSYSTEMS 61

Goals and Guidelines, 16
Selection of Research Sites, 64
  Infertile New World Moist Lowland Forest, 65; Fertile New World Moist
  Lowland Forest, 66; Southeast Asian Lowland Forest, 67; New World
  Deciduous Forest, 67
Objectives of the Tropical Ecosystem Research Project, 68
Recommended Subprojects, 70
  Water and Nutrient Cycling, 71; Ecosystem Energetics, 72;
  Physiological Plant Ecology, 74; Herbivory, 78; Higher-Order
  Food Webs, 80; Dynamics of Microhabitats and Patches, 81; Soils
  Research, 83

5 TROPICAL AQUATIC SYSTEMS   85

   Rivers, 86
      Very Large River Systems, 86; Smaller River Systems, 90
   Lakes, 90
      Closed-Basin Lakes, 91; Lakes with High Levels of Endemism, 92;
      Lakes Likely To Be Altered by Development, 93
   Wetlands, 93
      Large Swamps, 93; Major Riverine Wetlands, 95; Peat Swamps, 97
   Precipitation, 97
   Summary of Priorities, 97

REFERENCES   99

CONTRIBUTORS   113

Preface

Destruction of tropical vegetation has attracted the attention of many national and international bodies, especially during the past two decades. If this destruction continues at its present rate until the twenty-first century, it will lead to alteration in the course of evolution world wide, to widespread human misery, and to loss of the very knowledge that might be used to moderate the other consequences. This knowledge is therefore not only of theoretical importance but has critical application of benefit to human welfare, both in the tropical countries themselves and throughout the world. If tropical research is allowed to continue at its present level, it will contribute only in a minor way to the solution of the human and scientific problems to which we have alluded.

If the situation is to be improved, we must find ways to accelerate the pace of fundamental research and to concentrate some of it along especially promising lines and in critical areas.

The Assembly of Life Sciences, National Research Council, established the Committee on Research Priorities in Tropical Biology in July 1977. The main task of the committee was to consider the kinds of research needed in terrestrial and freshwater tropical biology and to determine those that should receive priority attention. We made an effort, therefore, to determine which questions could be addressed only in the tropics or could be addressed more appropriately there than elsewhere. We attempted to choose those questions that were of the most general character and to select them both because of their scientific interest and because the answers to them appeared to us to have great potential importance for improving the quality of human life. To provide as firm a base as possible for the determination of the urgency of particular issues, we commissioned a study of the rates of conversion of tropical forests world wide by Norman Myers (NRC, 1980).

We think of our program as one that will be undertaken cooperatively by many nations for their own benefit. We have concentrated on basic scientific studies that we believe are essential for both theoretical and practical reasons.

The committee held two meetings in St. Louis, Missouri (March and July 1978), and one meeting at the Instituto Nacional de Investigaciones sobre Recursos Bióticos, Xalapa, Veracruz, Mexico (May 1979).

Many persons assisted the committee in preparing this report, and we are grateful to them. Special thanks are due to persons who were guests at committee or panel meetings and participated in our discussions; they are named below. Others assisted the committee by furnishing information or reviewing parts of the draft of the report; they are named in the Appendix. Those who assisted at our meetings are the following:

JAMES C. ARONSON, St. Louis, Missouri
ALFREDO AND ALICIA BARRERA, Instituto Nacional de Investigaciones sobre Recursos Bióticos (INIREB), Xalapa, Veracruz, Mexico
VERNON H. HEYWOOD, Department of Botany, Plant Sciences Laboratory, The University of Reading, Whiteknights, Reading, England
EDUARDO JOHNSON, INIREB, Xalapa, Veracruz, Mexico
THOMAS E. LOVEJOY, Vice-President for Science, World Wildlife Fund—U.S., Washington, D.C.
NORMAN MYERS, Nairobi, Kenya
RAFAEL HERNÁNDEZ OCHOA, Governor of Veracruz, Xalapa, Mexico
JOSÉ SARUKHÁN, Departamento de Botánica, Instituto de Biología, Universidad Nacional Autónoma de México, Mexico
JEFFREY RICHEY, College of Fisheries, University of Washington, Seattle
S. H. SOHMER, Department of Botany, University of Wisconsin, La Crosse
GARY T. TOENNIESSEN, Agricultural Sciences Division, Rockefeller Foundation, New York

We are aware that a number of initiatives that have been undertaken throughout the world are directly related to ours. The immense amount of work that has been done in connection with the development of the technology of converting tropical forest for human benefit, for example, is of great interest to tropical biologists. Within the United States,

Preface xi

a Federal Interagency Task Force on Tropical Forests drafted a national policy on tropical forests and submitted it to the President in April 1980. This document contains an overview of U.S. and international efforts in the tropics and we have not attempted to repeat such a listing here. We have, of course, referred to the activities of particular agencies when these have been directly related to the research topics being discussed.

We have attempted in this report to fulfill our original mandate, namely, to identify research projects in tropical biology that are likely to yield results of great scientific and societal importance and to devise strategies whereby these might be carried out. We have not tried to cover all fields of tropical biology in an exhaustive listing; the report *Fragile Ecosystems* (Farnworth and Golley, 1974) provides a good catalog of important research topics for the neotropics. In general, we have concentrated on the kinds of basic scientific studies that we believe are essential for both theoretical and practical reasons rather than on the development of appropriate technologies for the tropics.

The report to the committee at large of its Panel on Human Ecology, which explored the interface of biological and humanistic concerns, made the case that, in addition to purely biological questions, there exists a wide spectrum of connected research topics, ranging from processes of destruction themselves to the complexities of human perception and behavior that provide the rationale for conversion; and, further, that the investigation of such topics is conducive to the discovery of viable alternatives capable of mitigating, halting, or reversing ongoing processes of environmental degradation. Thus the report recommended that multidisciplinary efforts unfold on at least two levels: (1) the investigation, in the tropical domain itself, of socioeconomic dysfunctions, such as those produced by transplanted models of development, and (2) the investigation of such dysfunctions in the broader perspective of national and international linkages: historical, political, social, and economic. Specifically, knowledge of the "systemic" characteristic of forest conversion demands research into the ways in which aboriginal, peasant, and urban-industrial societies perceive and relate to the tropical milieux and regulate their use.

The committee was convinced of the critical importance of the social concerns highlighted by the Panel on Human Ecology, believing that they ought to permeate the body of this report. As to the specific research topics that address the worldwide phenomena leading so inexorably to ecological instability in the tropics, the committee, while not analyzing them here, recommends interdisciplinary attention to these topics as a matter of highest priority. The same may be said of anthropological studies in general, which are of critical importance but lie outside our scope. For them, as for many other fields of study that we have not considered specifically, including such areas as the marine sciences and paleontology, the kinds of studies we have recommended are of importance, and the participation of tropical biologists is essential.

As would have been expected with any group of people with such diverse backgrounds, not all members of the committee agree with every viewpoint expressed here. Nonetheless, we have avoided positions so weak that they would be ineffectual and have refrained from cataloging all possible alternatives and actions with respect to a particular question. We hope that our recommendations will have a positive effect on the development of the vast and critically important fields that we have considered, and we will continue to work individually in the future to advance this aim.

Committee on Research Priorities
   in Tropical Biology

1 Summary and Recommendations

The tropical forests of the world are being altered so rapidly that most will not exist in their present form by the close of the century. The only extensive areas of undisturbed forest that probably will remain at that time, chiefly those in the western Brazilian Amazonia and in Central Africa, will probably persist for a few more decades.

Although the conversion of tropical forest often results in immediate economic gain, and some tropical lands have been converted to timber plantations that are productive on a long-term basis, systems that lead to the sustained productivity of most tropical soils have not been achieved with existing technology. Consequently, the likelihood of instability, both ecological and human, is increased dramatically as the forests are altered and eliminated.

Scientific knowledge about tropical ecosystems is extremely incomplete. It is estimated that a minimum of 3 million kinds of organisms occur in the tropics and that only about a sixth of these are known to science. The great majority of tropical organisms are therefore unknown. Moreover, there are few data pertaining to the functioning of aquatic and terrestrial ecosystems.

Virtually no knowledge is available to apply to the solution of many ecological problems related to those ecosystems. Without such knowledge it will be impossible to construct ecologically sound systems capable of supporting the numbers of people living in the tropics, to say nothing of improving the condition of those people. The problem is compounded by the fact that great differences occur from place to place throughout the tropics, even in areas occupied by vegetation types that are similar in appearance.

Given the scope and urgency of the situation, it is necessary to make hard choices among the problems to be studied and the particular areas that should receive attention. In making these choices, the committee utilized the following criteria:

- Urgency derived from imminent drastic change that might altogether preclude future study of those ecological systems or biological interactions.
- Potential scientific significance to the field of biology and its supporting disciplines of the study of particular systems or interactions, especially in the light of their applicability to human welfare.

In view of these considerations, the scientific and societal reasons for an accelerated research effort in the tropics of the world during the last few decades when such efforts will be possible seem compelling. There is an extreme urgency in setting priorities for research in tropical biology, not to displace any existing projects but to amplify the whole effort. Similarly, there needs to be a high priority placed on the detailed, multidisciplinary study of selected aboriginal peoples, whose way of life is changing very rapidly. Biologists should participate extensively in these studies when they are organized, but the formulation of the plan of action lies outside the scope of our committee. For the area of tropical biology itself, we offer the following recommendations, which are not necessarily listed in order of priority.

BIOLOGICAL INVENTORY

The international effort in completing an inventory of tropical organisms should be greatly accelerated, especially during the next 25 years. Our specific recommendations are as follows:

- Greatly accelerate the pace of biological inventory in the tropics, combining regional studies of relatively well-known groups or those of economic importance with detailed local inventories of others.
- Take steps to increase the pool of taxonomists studying tropical organisms from its present level of about 1,500 people world wide to four or five times as many, emphasizing the development of local institutions and the support of locally trained personnel in the tropical countries.
- Emphasize collecting specimens, in part by "unusual" methods,

Summary and Recommendations

such as freeze drying or preservation in fluids.
- Increase the conservation of genetic diversity in reserves and through the systematic organization of collections in botanical gardens, zoos, seed banks, tissue-culture banks, and whatever other means may be found suitable.
- Emphasize the biosystematics and evolutionary biology of a representative sample of tropical organisms, and then investigate these selected organisms by comprehensive multidisciplinary approach.
- Give priority to areas containing the richest and most highly endemic biota, the least-known biota, and biota in immediate danger of extinction. We suggest as areas for emphasis during the next 5 to 10 years the coastal forests of Ecuador, coastal southern Bahia and Espírito Santo in Brazil, eastern and southern Brazilian Amazonia, western and southern Cameroon and adjacent parts of Nigeria and Gabon, Hawaii, Madagascar, Sri Lanka, Borneo, Celebes, New Caledonia, and certain forested areas in Tanzania and adjacent Kenya.

TROPICAL ECOSYSTEM STUDIES

Tropical ecosystems should be investigated in depth at places selected because they are representative, diverse, and capable of experimental manipulation and because of scientific and societal importance. These studies should investigate both natural and experimentally manipulated ecosystems and should emphasize solutions to problems in areas of general ecological interest; the problems should include mineral flows and cycling, species richness, ecosystem energetics, and ecosystem stability. Sites selected for such detailed study include New World moist rain-forest sites having soils of comparatively high (La Selva, Costa Rica) and low (*terra firme* of the central Amazon Basin) fertility, a site in New World tropical deciduous forest (Chamela, Jalisco, Mexico), and a site in moist tropical forest in Asian lowland (Mulu, Sarawak). Other areas of emphasis should include Puerto Rico and Hawaii as examples of tropical island ecosystems and of tropical grasslands and savannas. The regional projects under United Nations Educational, Scientific and Cultural Organization—Man and the Biosphere (UNESCO–MAB) Project 1 and their associated training activities should be developed in cooperation with these recommended sites.

Among the objectives of the tropical ecosystem research project would be the establishment and continued monitoring of key parameters of the physical and biological environment at each site: water and nutrient cycling, ecosystem energetics, seasonal rhythms of leaf and fruit production, physiological plant ecology, herbivory, food webs, and the dynamics of microhabitats and patches. We expect that knowledge gained at each of these sites would interact with knowledge gained at all the others and that the result would be rich insights into the functioning of tropical ecosystems in general. The studies should help in establishing the factors involved in creating continuously productive ecosystems in the tropics and in indicating base-line conditions in many areas of ecology and population biology.

In the area of soils research, we already know the broad correlations between soil and vegetation in the seasonal tropics and in the humid tropics of Asia and parts of Africa, though more needs to be done to corroborate the patterns established in South America. The dynamic interaction between forest and soil in different climates and on different soils is still poorly understood. Detailed study of the movement of nutrient and trace-element ions in solution through plants, fauna, and soil are required; comparisons between primary forest and various forms of converted land will provide an essential basis for developing sustainable forms of land use. Such studies should be concentrated, though not exclusively carried out, at the major ecosystem sites defined in this report.

In the field of tropical physiological plant ecology, we recommend the establishment of a comprehensive laboratory in Puerto Rico, which would become a center for research and training in the field. The creation of such a laboratory appears to be critical for the establishment of a field of investigation that is poorly developed in the tropics. Such investigation appears to offer a great deal in both scientific and societal rewards.

STUDIES OF TROPICAL AQUATIC SYSTEMS

We believe that tropical freshwater systems should be studied much more intensively than at present in view of their scientific and economic importance.

Rivers

We suggest that immediate efforts be made to support and further strengthen the ongoing study of the Amazon and Orinoco rivers and their main branches. The Zaïre should also have high priority for study. Studies of the structure and functioning of these rivers should include water chemistry and nutrient processing, plankton, fishes, bottom fauna, key vertebrates, and land–water interactions. As for smaller rivers, priority attention should be given to the Musi River in Sumatra and the Purari River in Papua New Guinea and other rivers similarly

Summary and Recommendations 5

threatened with change. Intensive study should be followed by long-term monitoring in all these rivers.

Lakes

We emphasize closed-basin lakes, lakes that contain major assemblages of endemic species, and lakes that are subject to change in the immediate future. A combination of these criteria has led us to the conclusion that the study of the following lakes is urgent: Lakes Valencia and Maracaibo in Venezuela, Lake Malawi in Africa, Lake Titicaca in South America, and the volcanic lakes in insular Southeast Asia.

Wetlands

Wetlands are of great ecological significance and are highly vulnerable to destruction. We place the highest priority on study of the Sudd, an extensive swamp of unique character maintained by the Nile. Major drainage projects have already been initiated in this swamp. The Pantanal of Mato Grosso in Brazil, which is largely dry land for part of the year, is even larger and perhaps has greater regional importance. It faces active development. At a somewhat lower priority among large tropical swamps, we place those of the Territorio Amapá in Brazil, those of Beni Department in Bolivia, and the Benguelu Swamp of Zambia, largely because they are less threatened in the immediate future (i.e., within 5 years). Of the major riverine wetlands, much of the *várzea* of the Amazon Basin and the delta and backwaters of the Orinoco appear to be immediately threatened; the backwaters of the Zaïre and Xingu appear to be less threatened. The extensive peat swamps of Southeast Asia demand attention because of their extent and because they are an important resource that is being exploited at an increasing rate.

Precipitation

Precipitation chemistry and amounts, which are relevant to the characteristics of both terrestrial and freshwater environments, ought to be monitored and studied, especially in the Amazon, Orinoco, and Mekong drainages.

MONITORING FOREST CONVERSION

National schemes for monitoring the rates of conversion of tropical moist forests and other tropical vegetation types should be encouraged and, when appropriate, aided by competent international bodies. See NRC (1980).

FIVE-YEAR SCHEDULE

During the 5-year period 1980–1985, we call for the following actions as matters of extreme importance for the attention of all the nations of the world:

- At least double, in constant dollars, the funds now devoted to biological inventory in the tropics.
- Increase by at least 50% the number of professional systematists engaged in studies of tropical organisms.
- Initiate operations in at least the basic four ecosystem sites mentioned, establish a center for the study of tropical plant physiological ecology, and complete the installation of their physical facilities. Complete the basic studies of mineral cycling and the basic biological and soils inventories at each ecosystem site.
- Initiate or expand major comprehensive studies of the structure and functioning of the Amazon, Orinoco, and Purari rivers and their major branches; of Lakes Valencia and Maracaibo; and of the Sudd, the Pantanal of Mato Grosso, the *várzea* of the Amazon Basin, and the delta and backwaters of the Orinoco. The basic 5-year studies we have outlined should be completed in all these cases well before 1990, and studies of other subjects should be initiated in the period 1986–1990.
- Fund national monitoring and international reporting of the rates of conversion of tropical vegetation types, especially tropical moist forest.

1 Exposé Succinct et Recommandations

Les forêts tropicales dans le monde subissent des transformations si rapides que la plupart d'entre elles n'existeront pas sous leur forme actuelle à la fin du siècle. Les seules régions étendues de forêts paisibles qui demeureront probablement à cette époque-là, principalement celles de l'Amazone brésilienne occidentale et de l'Afrique centrale, persisteront, en toute probabilité, pendant quelques décennies de plus.

Bien que la conversion de la forêt tropicale ait souvent pour conséquence un avantage économique immédiat, et que quelques terres tropicales ont été converties en plantations productrices de bois à longue échéance, les systèmes entraînant une productivité constante pour la plupart des sols tropicaux n'ont pas été menés à bien avec la technologie existante. En conséquence, les chances d'instabilité, à la fois écologique et humaine, se trouvent gravement accrues avec la transformation et l'élimination graduelles des forêts.

La connaissance scientifique des systèmes écologiques tropicaux est extrêmement incomplète. On estime que trois millions au moins de variétés d'organismes se trouvent dans les tropiques et que la science n'en connaît qu'un sixième environ. La grande majorité des organismes tropicaux est donc inconnue. Il n'y a virtuellement aucune connaissance disponible que l'on pourrait appliquer à la solution des nombreux problèmes écologiques ayant un rapport avec ces systèmes écologiques. Sans cette connaissance il sera impossible de construire des systèmes écologiquement solides, capables de subvenir aux besoins des habitants des tropiques, encore moins d'améliorer leur condition.

Le problème est compliqué par le fait que de grandes différences surgissent d'un endroit à l'autre dans les tropiques, même dans les regions couvertes de végétations de types apparemment semblables.

Etant donné l'envergure et l'urgence de la situation, il est nécessaire de faire un choix difficile entre les problèmes à étudier et les régions qui méritent une attention particulière. En faisant ce choix, le Comité s'est basé sur les critères suivants:

• L'urgence découlant d'un changement draconien imminent qui pourrait entièrement empêcher une étude future de tels systèmes écologiques ou interactions biologiques.

• La signification scientifique potentielle, dans le domaine de la biologie et des disciplines connexes, de l'étude des systèmes ou interactions particuliers, surtout à la lumière de leur applicabilité au bien-être humain.

En vue de ces considérations, les raisons scientifiques et sociétales pour un effort de recherche accéléré dans les tropiques du monde paraissent contraignantes. Il est extrêmement urgent de fixer les priorités pour la recherche en biologie tropicale, pour amplifier l'effort dans son ensemble sans évincer aucun des projets existants. De même, un besoin prioritaire se fait sentir dans l'étude detaillée, multidisciplinaire de peuples aborigènes sélectionnés dont le mode de vie est en train de changer très rapidement. Des biologistes devraient participer largement à ces études une fois mises en place, mais la formulation du plan d'action même n'est pas de la compétence de notre Comité. Quant à la zone même de biologie tropicale, nous faisons les recommandations suivantes, pas nécessairement dans l'ordre de leurs priorités.

INVENTAIRE BIOLOGIQUE

L'effort international de dresser un inventaire des organismes tropicaux devrait être fortement accéléré, sourtout dans les 25 prochaines années. Ci-après nos recommandations précises:

• Accélérer fortement l'allure de l'inventaire biologique dans les tropiques, combinant les études régionales de groupes relativement bien connus ou économiquement importants à d'autres inventaires locaux détaillés.

• Prendre des mesures pour augmenter l'équipe des taxonomistes étudiant les organismes tropicaux et la porter de son niveau mondial actuel de 1500 personnes à quatre ou cinq fois plus, en mettant l'accent sur le développement des institutions locales et l'assistance d'un personnel local formé sur place.

- Mettre l'accent sur la collection des spécimens, en partie par des méthodes "inaccoutumées" telles que la congélation sèche ou la préservation dans des fluides.
- Augmenter la conservation de la diversité génétique dans des réserves et, par l'organisation systématique de collections dans des jardins botaniques, des zoos, des banques de graines, des banques de culture de tissus, et tous autres moyens qu'on trouverait appropriés.
- Mettre l'action sur la classification biologique et la biologie évolutionnaire d'un échantillon représentatif d'organismes tropicaux, puis examiner ces organismes sélectionnés en détail grâce à des travaux multidisciplinaires étendus.
- Accorder la priorité aux régions contenant les spécimens biotiques les plus hautement endémiques, les moins connus, et ceux en danger d'extinction immédiate. Nous suggérons de porter une attention spéciale à des régions telles que les forêts côtières de l'Equateur, les côtes méridionales de Bahia et d'Espirito Santo au Brésil, l'Est et le Sud de l'Amazone brésilienne, l'Ouest et le Sud du Cameroun, et les parties adjacentes du Nigéria et du Gabon, Hawaï, Madagascar, Sri Lanka, Bornéo, les Célèbes, la Nouvelle-Calédonie, et certaines régions forestières de la Tanzanie et du Kenya adjacent.

ETUDES DES SYSTÈMES ECOLOGIQUES TROPICAUX

Les systèmes écologiques tropicaux devraient être examinés en profondeur dans des régions choisies pour leur représentativité, leur diversité et leur capacité de manipulation expérimentale ainsi que pour leur importance scientifique et sociétale. Ces études devraient examiner à la fois les systèmes écologiques naturels et manipulés expérimentalement et devraient mettre l'accent sur les solutions aux problèmes dans des régions d'intérêt écologique général; les problèmes devraient inclure les courants et mouvements cycliques minéraux, la richesse des espèces, l'energétique des systèmes écologiques et leur stabilité. Les cites sélectionnés pour une telle étude détaillée comprennent les sites des forêts humides du Nouveau Monde dont la fertilité du sol est relativement élevée (La Selva, Costa Rica) et réduite (*terra firme* du bassin central de l'Amazone), un site dans la forêt tropicale caduque du Nouveau Monde (Chamela, Jalisco, Le Mexique) et un site dans la forêt tropicale humide des plaines asiatiques (Moulou, Sarawak). D'autres régions d'importance devraient comprendre Porto-Rico et Hawaï servant d'exemples de systèmes écologiques tropicaux insulaires et de prairies et savanes tropicales. Les projets régionaux sous le Projet 1 de l'UNESCO–MAB et leurs activités associées de formation devraient être mis en valeur en coopération avec ces sites recommandés.

Parmi les objectifs du projet de recherche dans les systèmes écologiques tropicaux on compterait l'établissement et le contrôle continu de paramètres clef de l'environnement physique et biologique sur chaque site, les mouvements cycliques de l'eau et de la nutrition, l'energétique des systèmes écologiques, les rythmes saisonniers de la production de feuillages et de fruits, l'écologie physiologique des plantes, l'herbivorie, les cycles naturels alimentaires, et la dynamique des microhabitats et des parcelles de terrain. Nous présumons qu'une action réciproque interviendrait entre la connaissance acquise dans chacun de ces sites et celle acquise dans tous les autres, et qu'il en résulterait une compréhension substantielle du fonctionnement des systèmes écologiques tropicaux en général. Ces études devraient aider à déterminer les facteurs qui militent dans la création de systèmes économiques constamment productifs dans les tropiques, et à définir les conditions fondamentales de la biologie écologique et humaine dans plusieurs régions.

Dans le domaine de la recherche des sols, nous savons déjà que de vastes corrélations existent entre le sol et la végétation dans les tropiques saisonniers et dans les tropiques humides de l'Asie et de certaines parties de l'Afrique, quoiqu'un effort supplémentaire soit nécessaire pour corroborer les tendances prévalant en Amérique du Sud. L'interaction dynamique entre la forêt et le sol sous des climats différents et sur des sols différents est encore mal comprise. Une étude détaillée du mouvement des ions nutritifs et des micro-éléments en solution dans les plantes, la faune et le sol, est nécessaire; des comparaisons entre des forêts primaires et des formes diverses de terre convertie fourniront une base essentielle pour développer des formes valables d'usage de la terre. Ces études devraient être concentrées, bien que non exclusivement exécutées, sur les sites principaux des systèmes écologiques définis dans ce rapport.

Dans le domaine de l'écologie tropicale et physiologique des plantes, nous recommandons l'établissement d'un vaste laboratoire à Porto-Rico qui deviendrait un centre de recherche et de formation dans ce domaine. La création de ce laboratoire semble être critique pour l'établissement d'un champ d'investigation peu developpé dans les tropiques. Cette investigation semble offrir la promesse de récompenses à la fois scientifiques et sociétales.

Exposé Succinct et Recommandations 11

ETUDES DES SYSTÈMES TROPICAUX AQUATIQUES

Nous pensons que les systèmes tropicaux d'eau douce devraient être étudiés beaucoup plus intensivement que dans le présent en raison de leur importance scientifique et économique.

Fleuves

Nous proposons que des efforts immédiats soient déployés pour soutenir et puis renforcer l'étude en cours des fleuves de l'Amazone et de l'Orénoque et de leurs bras principaux. Le Zaïre devrait être étudié en toute priorité. L'étude de ces fleuves devraient comprendre la chimie aquatique et le traitement nutritif, le plancton, les poissons, la faune des profondeurs, les vertébrés clef et les interactions sol-eau. Quant aux plus petits fleuves, une attention prioritaire devrait être portée au Musi à Sumatra et au Purari en Nouvelle-Guinée et à d'autres fleuves également menacés de transformation. Une étude intensive devrait être suivie d'un contrôle à long terme dans tous ces fleuves.

Lacs

Nous mettons l'accent sur les lacs à circuit fermé, les lacs qui contiennent des assemblages importants d'espèces endémiques, et les lacs appéls à se transformer dans un avenir immédiat. Une combinaison de ces critères nous a conduits à conclure que l'étude des lacs suivants est urgente: les lacs Valencia et Maracaïbo au Venezuela, le lac Malawi en Afrique, le lac Titicaca en Amérique du Sud, et les lacs volcaniques de l'Asie insulaire du Sud-Est.

Terres Marécageuses

Les terres marécageuses sont d'une grande importance écologique et tres vulnérables à la destruction. Nous accordons la plus haute priorité à l'étude du Sudd, un marécage très étendu à caractère unique alimenté par le Nil. Des projets majeurs de drainage ont été déjà mis en oeuvre dans ce marécage. Le Pantanal du Mato Grosso au Brésil, qui est un terrain sec une partie de l'année, est même plus grand et a probablement une plus grande importance régionale. Il subit une évolution active. En moindre priorité, nous plaçons dans une certaine mesure parmiles grands marécages tropicaux, ceux du Territoire Amapa au Brésil, ceux du Département Beni en Bolivie, et le marécage Benguelou en Zambie, surtout parce qu'ils sont moins menacés dans l'avenir immédiat (c'est-à-dire dans les 5 ans). Parmi les principales terres marécageuses fluviales, la plus grande partie du *várzea* de bassin de l'Amazone et du delta et des bras de décharge de l'Orénoque semble menacée dans l'immédiat; les bras de décharge du Zaïre et du Xingu semblent être moins menacés. Les tourbières étendues de l'Asie du Sud-Est retiennent l'attention à cause de leur étendue et parce qu'elles constituent une ressource importante qui est en train d'être exploitée à une vitesse accrue.

Précipitations

La chimie et la quantité de précipitation qui sont propres aux environnements à la fois terrestres et d'eau douce devraient être contrôlées et étudiées, surtout dans les bassins hydrographiques de l'Amazone, de l'Orénoque et du Mékong.

CONTRÔLE DE LA CONVERSION FORESTIÈRE

Des plans nationaux pour contrôler le train de conversion des forêts humides tropicaux et d'autres types de végétation tropicale devraient être encouragés et, en temps opportun, assistés par des organisations internationales compétentes. Voir NRC (1980).

CALENDRIER QUINQUENNAL

Pendant la période de 5 ans, 1980–1985, nous demandons les mesures suivantes en tant que matières d'importance primordiale à porter à l'attention de toutes les nations du monde:

• Doubler au moins, en dollars constants, les fonds consacrés à l'inventaire biologique dans les tropiques.
• Augmenter de 50% au moins le nombre de classificateurs professionnels engagés dans l'étude des organismes tropicaux.
• Mettre en oeuvre des opérations dans les quatre sites typiques à systèmes écologiques précités, installer un centre pour l'étude de l'écologie physiologique des plantes tropicales, et compléter l'installation de leurs facilités physiques. Terminer les études essentielles des mouvements cycliques minéraux et des inventaires biologiques et de sols fondamentaux dans chaque site à système écologique.
• Introduire ou développer des études majeures de grande portée des fleuves de l'Amazone, de l'Orénoque et du Purari et deleurs bras principaux, des lacs Valencia et Maracaïbo et du Sudd, du Pantanal

dans le Mato Grosso, du *várzea* dans le bassin de l'Amazone et du delta et des bras de décharge de l'Orénoque. Les études quinquennales essentielles que nous avons esquissées devraient être terminées, dans tous les cas, bien avant 1990, et des études d'autres sujets devraient être instaurées dans la période 1986–1990.

• Consolider le contrôle national et les transmissions internationales des trains de conversion des types de végétation tropicale, surtout dans la forêt humide tropicale.

1 Sumário e Recomendacoes

As florestas tropicais do mundo têm sido alteradas tão ràpidamente que, no final do século, muitas não terão a sua forma presente. As únicas áreas florestais extensas que provàvelmente permanecerão inalteráveis até essa altura, serão principalmente o Oeste da Amazónia Brasileira e na África Central, que provàvelmente persistirão por mais algumas decadas.

No entanto a transformação de florestas tropicais resulta muitas vezes num imediato lucro económico, e algumas terras tropicais têm sido convertidas em plantações fornecedoras de madeiras que são produtivas a longo termo, sistemas que conduzem a uma produtividade constante em muitos solos tropicais, ainda não conseguidas com a tecnologia existente. Por conseguinte, a probabilidade de instabilidade, ecológica e humana, aumenta dramàticamente a medida em que as florestas são alteradas e eliminadas.

O conhecimento científico sobre sistemas ecológicos tropicais está extremamente incomplecto. Calcula-se que num mínimo de tres milhões de espécies orgânicas que ocorrem nos trópicos, apenas um sexto é cientificamente conhecido. A grande maioria de organismos tropicais são por consequência desconhecidos. Não existe virtualmente qualquer conhecimento para aplicar à solução de muitos problemas ecológicos relacionados com esses sistemas. Sem tal conhecimento será impossível de ecològicamente construir sistemas precisos, capazes de ajudar as populações dos trópicos, para não dizer o melhoramento das suas condições de vida. O problema complica-se pelo facto de que

Sumário e Recomendações

grandes diferenças ocorrem de lugar para lugar ao longo dos trópicos, mesmo em areas ocupadas por tipos de vegetação em aparência semelhantes.

Dada a natureza e a urgência da situação, torna-se necessário fazer escolhas difíceis entre os problemas a serem estudados e particularmente as áreas que devem receber atensão. Para fazer estas escolhas o Comité utilizou o seguinte critério:

- A urgência decurrente da mudança drástica eminente que poderá impedir por completo o futuro estudo dos sistemas ecológicos ou das interações biológicas.
- Significado científico potencial para o campo da biologia e disciplinas colaterais do estudo de sistemas ou interações particulares especialmente à luz da sua aplicação para o bem estar humano.

Em vista destas considerações as razões científicas e sociais para um esforço de pesquisa acelerado no mundo tropical, durante as últimas decadas, parecem ser imperativas quando sejam possíveis. É extremamente urgente estabelecer prioridade para a pesquisa no campo da biologia tropical, não diferindo os projectos existentes mas aplicar todo o esforço. Do mesmo modo é necessário dar a maior e mais detalhada prioridade ao estudo múltiplo de povos aborigenes seleccionados, cujo modo de vida está a mudar muito ràpidamente. Os biólogos deveriam participar devidamente nestes estudos quando forem organizados, mas a formulação do plano de acção deveria ser exterior ao nosso Comité. No respeitante à biologia tropical em si apresentamos as seguintes recomendações, que não são necessariamente indicadas pela ordem de prioridade.

INVENTÁRIO BIOLÓGICO

O esforço internacional para a conclusão do inventário de organismos tropicais deveria ser largamente acelerado, especialmente durante os próximos 25 anos. As nossas recomendações específicas são as seguintes:

- Acelerar intensivamente o ritmo de inventário biológico nos trópicos, combinando os estudos regionais de grupos relativamente bem conhecidos, com os de importância económica e com detalhados inventários locais de outras espécies.
- Tomar medidas para incrementar os estudos de taxonomia sobre organismos tropicais por forma a elevá-los do seu presente nível de

cerca de 1.500 pessoas à escala mundial para o quadruplo ou quintuplo, por forma a salientar o desenvolvimento de instituições locais e o apoio de pessoal local especializado, ou seja nos próprios paizes tropicais.

• Incrementar a recolha de espécies, em parte através, de meios invulgares como congelamento a seco ou a preservação em fluidos.

• Aumentar a conservação de diversas espécies genéticas em reservas e através de organização sistemática de colecções em jardins botânicos, jardins zoológicos, reservas de sementes, cultura de tecidos e todos os outros meios que sejam considerados adequados.

• Incrementar a sistemática e evolução biológicas de uma amostragem representativa de organismos tropicais, e depois fazer a investigação desses organismos seleccionados através de meios multilaterais.

• Dar prioridade a áreas contendo as mais ricas e elevadas biologias endémicas, as menos conhecidas e as que estão em perigo de imediata extinsão. Sugerimos como áreas de importância, durante os próximos 5 a 10 anos, as áreas florestais costeiras do Equador; as zonas costeiras do Sul da Baia e do Espirito Santo no Brasil; assim como as zonas de Leste e Sul da Amazónia, também no Brasil; as zonas Ocidental e Austral dos Camarões e as áreas adjacentes da Nigéria e do Gabão; Hawaii; Madagáscar; Sri Lanca; Borneo; Celebes; a Nova Caledónia e certas áreas florestais na Tanzânia e no paiz vizinho, o Quénia.

SISTEMAS DE ESTUDOS ECOLÓGICOS TROPICAIS

Os sistemas tropicais ecológicos devem ser investigados profundamente nos lugares seleccionados porque são representativos, variados, e possíveis de ser manipulados devido à sua importância científica e social. Estes estudos devem investigar tanto os sistemas ecológicos naturais como os que são manipulados experimentalmente devendo também realçar soluções para os problemas em áreas de interesse ecológico geral; os problemas devem incluir correntes minerais e cíclicas, riqueza de espécies, a energética e a estabilidade dos sistemas ecológicos. Lugares seleccionados para tal detalhado estudo incluem as florestas húmidas do Novo Mundo, lugares tendo solos comparativamente elevados (Le Selva, Costa Rica) e baixos (*terra firme* da Bacia Central da Amazónia) férteis, uma floresta tropical moribunda no Novo Mundo (Chamela, Jalisco, México) e um lugar na floresta tropical húmida da terra-baixa Asiática (Mulu, Sarawak). Outras áreas de realce devem incluir Porto Rico e Hawaii como exemplos de ilhas de sistemas ecológicos tropicais e de terrenos de capim e savanas tropicais. Os projectos regionais sob o Projecto 1 da UNESCO–MAB e sob as suas

Sumário e Recomendações

actividades associadas de formação, devem ser desenvolvidas em cooperação com os lugares recomendados.

Entre os objectivos do projecto de pesquisa do sistema ecológico tropical deve estar o estabelecimento e o continuo control das chaves paramétricas do meio-hambiente físico e biológico local: água e ciclos de nutrição; sistemas ecológicos energéticos; ritmos de estações de produção de folhas e frutos; fisiologia ecológica da planta; hervanoso; veios de nutrição; e as dinâmicas de micro-habitates e remedeios. Esperamos que o conhecimento adquirido em cada um destes lugares seja correlacionado com o conhecimento ganho em todos os outros, e assim o resultado venha a ser rico em inteligência no funcionamento dos sistemas ecológicos tropicais em geral. Os estudos devem ajudar a estabelecer os factores envolvidos a crear continuamente sistemas ecológicos productivos nos trópicos e a indicar condições de base em muitas áreas ecológicas e a biologia da população.

Na região de solos de pesquisa já se conhecem amplas correlações entre solos e vegetação nos trópicos de estações regulares, nos trópicos húmidos da Ásia e partes de África, é contudo necessário fazer muito mais para corroborar os padrões estabelecidos na América do Sul. A reciprocidade dinâmica entre florestas e solo em climas e solos diferentes continua a ser pouco compreendida. Um estudo detalhado do movimento nutritivo e vestígios de elementos de iodo em solução entre as plantas, fauna e solo são necessários; comparações entre florestas principais e várias formas de solos possíveis de transformação, fornecerão uma base issencial para formas de desenvolvimento constante da utilização das terras. Tais estudos devem ser concentrados, ainda que não exclusivamente conduzidos, no principal sistema ecológico dos lugares defenidos neste relatório.

No campo ecológico da fisiologia da planta tropical, recomendamos o estabelecimento de um laboratório complecto em Porto Rico, o qual virá a ser um centro de pesquisa e de formação em campo. A criação de tal laboratório parece ser decisiva para o estabelecimento de um campo de investigação que está pouco desenvolvido nos trópicos. Tal investigação parece oferecer bastante recompensa no campo científico e social.

ESTUDOS DE SISTEMAS AQUÁTICOS TROPICAIS

Acreditamos que os sistemas de água doce tropicais devem ser estudados muitos mais intensivamente do que o são presentemente, devido à sua importância científica e económica.

Rios

Sugerimos que sejam realizados esforços imediatos para suportar e reenforçar o estudo em marcha nos rios Amazonas e Orinoco e seus maiores afluentes. O Zaire também deve ter grande prioridade de estudo. Os estudos destes rios devem incluir a química da água e o processo de nutrição, planton, peixes, faunas das profundidades, principais vertebrados e interações de água e solo. Assim, para os rios menores deve ser dada prioridade ao rio Musi na Sumatra e ao rio Purari em Papua Nova Guiné e outros rios similares ameaçados de modificação. Um estudo intenso deve ser seguido por uma supervisão a longo prazo, em todos estes rios.

Lagos

Acentuamos chegadas-bacias fluviais, lagos que contêm maiores agrupamentos de espécies endémicas e lagos que estão sugeitos a modificações no futuro próximo. A combinação deste critério levounos a conclusão de que é urgente o estudo dos seguintes lagos: Lagos Valência e Maracaibo na Venezuela, Lago Malawi em África, Lago Titicaca na América do Sul, e os lagos vulcanicos no Insular Sudeste Asiático, cujo estudo deve ser cedo começado.

Terras Alagadiças

As terras alagadiças são de grande significado ecológico e altamente vulneráveis à destruição. Damos a maior prioridade ao estudo do Sudd, um muito extenso pântano de carácter único, mantido pelo Nilo. Grandes projectos de drenagem foram já aí iniciados. Ainda maior é o Pantanal de Mato Grosso no Brasil, o qual se encontra seco durante a maior parte do ano, e talvez tenha maior importância regional, enfrentando activo desenvolvimento. Com um pouco menos de prioridade, entre os maiores pântanos tropicais, colocamos os do território Amapá no Brasil, os do Departamento Beni na Bolívia e o pântano Benguelu da Zambia, em grande parte porque estão menos ameaçados, no próximo periodo de 5 anos. Das mais importantes margens de rios em terras alagadiças, a grande parte da *várzea* da Bacia do Amazonas e do Delta e a águas estagnadas de Orinoco, parecem estar imediatamente ameaçadas; as águas estagnadas do Zaire e Xingu parecem estar menos ameaçadas. Os extensos pântanos de turfas* do Sudeste Asiático mere-

*Ou musgos.

Sumário e Recomendações

cem atensão até porque são uma importante fonte que tem sido explorada a um ritmo acelerado.

Precipitação

A química e a quantidade de precipitação são caracteristicamente significativas no meio-hambiente terrestre e de água doce, devendo ser controlados e estudados especialmente na Amazónia, Orinoco e nas drenagens do Mekong.

CONTROL DA TRANSFORMAÇÃO DE FLORESTAS

Esquemas nacionais para o control da percentagem de transformação de florestas húmidas tropicais noutros tipos de vegetação tropical, devem ser encorajados e, quando for apropriado, ajudados por competentes corpos internacionais. Ver NRC (1980).

PROGRAMA DE 5 ANOS

Durante o periodo de 5 anos, 1980–1985, chamamos a atensão para as seguintes operações, que são assuntos de extrema importancia para a atensão de todas as nações do mundo:

- No mínimo duplicar em dolar constante, fundos dedicados ao inventário biológico dos trópicos.
- Aumentar no minimo de 50%, o número de sistematologistas empenhados no estudo de organismos tropicais.
- Iniciar operações, pelo menos nos quatro lugares mensionados de básicos sistemas ecológicos, criar um centro para o estudo físio-ecológico da planta tropical e complectar as instalações respectivas, complectar os estudos básicos de ciclos minerais e os inventários de base biológica e dos solos em cada local de sistema ecológico.
- Iniciar ou expandir os principais estudos exaustivos dos rios Amazonas, Orinoco e Purari e seus maiores afluentes; dos lagos Valência e Maracaibo; e do Sudd, o Pantanal de Mato Grosso, a *várzea* da Bacia do Amazonas e do delta e terras pantanosas do Orinoco. Os básicos 5 anos de estudo que apontamos devem ser concluidos em todos estes casos bem antes de 1990, e estudos sobre outras matérias devem ser iniciadas no periodo de 1986–1990.
- Fundo nacional de control e informação internacional do grau de transformação de tipos de vegetação tropical, especialmente de florestas húmidas tropicais.

1 Resumen y Recomendaciones

Por el paso acelerado al cual se alteran los bosques tropicales del mundo la gran mayoría de estos no sobrevivirán este siglo en su forma actual. Las únicas áreas selváticas vírgenes que permanecerían entonces, principalmente las del Amazonas occidental en Brazil y del Africa central, probablemente perdurarían algunas décadas más.

Aunque la transformación de los bosques tropicales resulta a menudo en una ganancia económica inmediata, y algunas tierras has sido convertidas en plantaciones de árboles que son productivas a largo plazo, los sistemas que llevan a una productividad continua de las tierras tropicales no se han logrado con la tecnología existente. Por lo tanto, se incrementa dramáticamente la probabilidad de inestabilidad ecológica y humana a medida que se alteran y eliminan estos bosques.

Los conocimientos científicos en torno a los ecosistemas tropicales son sumamente incompletos. Se calcula que existe un mínimo de tres millones de diferentes clases de organismos en el trópico pero de estos se conoce científicamente solamente una sexta parte. La gran mayoría de estos organismos tropicales permanecen desconocidos. No existe virtualmente ninguna información para solucionar los numerosos problemas ecológicos relacionados a estos ecosistemas. Sin estos conocimientos será imposible crear los sistemas ecológicos adecuados para mantener el número de habitantes del trópico, y menos todavía pensar en mejorar su condición de vida. El problema se complica por las fuertes diferencias que se encuentran de un lugar a otro en las tierras tropicales, aún en las extensiones cubiertas con una vegetación de apariencia semejante.

Por la magnitud y la urgencia de la situación debe efectuarse una

Resumen y Recomendaciones 21

penosa selección entre los problemas por estudiar y las áreas especiales que merecen atención. Al hacer esta selección, el Comité se basó sobre los siguientes criterios:

• La urgencia derivada del cambio drástico inminente que puede precluir cualquier estudio futuro de esos sistemas ecológicos o interacciones biológicas.
• El significado científico latente del estudio de sistemas o interacciones particulares, al campo de la biología y sus disciplinas sustentadoras, especialmente en relación a su pertinencia al bienestar humano.

En vista de estas consideraciones, parecen apremiantes las razones científicas y sociales para realizar un esfuerzo e investigación acelerado en las regiones tropicales del mundo en las pocas décadas en que serán todavía posibles tales esfuerzos. Existe una urgencia obligatoria en el establecimiento de prioridades en la investigación de la biología tropical no para suplantar los proyectos existentes sino para amplificar el esfuerzo total. De igual manera, se necesita dar una alta prioridad al estudio detallado y multi-disciplinario de ciertos pueblos de aborígenes cuya forma de vida cambia rápidamente. Los biólogos deben participar extensamente en la organización de estos estudios pero queda fuera de nuestro Comité la formulación del plan de acción. En cuanto al campo de la biología tropical en sí, ofrecemos las siguientes recomendaciones que no se encuentran necesariamente en orden de prioridad.

INVENTARIO BIOLÓGICO

El esfuerzo internacional en completar un inventario biológico debe acelerarse fuertemente, especialmente en los próximos 25 años. Nuestras recomendaciones específicas son las siguientes:

• Acelerar fuertemente el paso del inventario biológico en los trópicos, combinando estudios regionales de grupos relativamente conocidos o también aquellos de importancia económica con los inventarios locales detallados de otros grupos.
• Tomar medidas para incrementar el nivel mundial actual de unos 1500 taxonomistas, que estudian los organismos tropicales, cuatro o cinco veces destacando el desarrollo de instituciones locales y el respaldo de personal adiestrado localmente en las regiones tropicales.
• Enfatizar la recopilación de especímenes, en parte con métodos poco comunes como la deshidratación por congelación y la preservación en liquidos.

- Desarrollar la conservación de la diversidad genética existente en tierras reservadas y a través de la organización sistemática de colecciones en parques botánicos, jardines zoológicos, bancos de semillas y de tejidos de cultivo y cualquier otro método adecuado.
- Destacar la biosistemática y la biología evolucionaria de un muestrario representativo de los organismos tropicales y luego, investigar estos organismos seleccionados con un amplio enfoque multidisciplinario.
- Darle prioridad a las regiones que poseen una biota endémica más notable y rica, a aquellos de biota menos conocida, y a las de biota en inminente peligro de extinción. Sugerimos como regiones que merecen énfasis en los próximos 5 a 10 años los bosques costaneros del Ecuador; la costa sur de Bahia y Espírito Santo en Brazil; el Amazonas al oriente y al sur del Brazil; el oeste y el sur del Camerún y partes adyacentes de Nigeria y Gabón; Hawaii; Madagascar; Sri Lanka; Borneo; Celebes; Nueva Caledonia; y ciertas regiones boscosas en Tanzanía y Kenya.

ESTUDIOS DE ECOSISTEMAS TROPICALES

Los ecosistemas tropicales deben investigarse en profundidad en los lugares seleccionados ya que son representativos, diversos y capaces de manipulación experimental y por su importancia científica y social. Estos estudios deben investigar los ecosistemas naturales y los que han sido manipulados experimentalmente y deben destacar las soluciones de problemas de interés ecológico general; los problemas deben incluir los ciclos y flujos de minerales, la riqueza de las especies, la energética del ecosistema y la estabilidad del ecosistema. Los lugares seleccionados para dichos estudios detallados incluyen los bosques tropicales húmedos del Nuevo Mundo con tierras de fertilidad relativamente alta (La Selva, Costa Rica) y baja (*terra firme* de la cuenca central del Amazonas), un lugar en los bosques deciduos del Nuevo Mundo (Chamela, Jalisco, Mexico) y un lugar del bosque tropical húmedo en tierras bajas asiáticas (Mulu, Sarawak). Otros lugares importantes deben incluir a Puerto Rico y Hawaii como ejemplares de ecosistemas tropicales de islas y de prados y savanas tropicales. Los proyectos regionales bajo el Proyecto I de la UNESCO–MAB y sus actividades de adiestramiento relacionadas deben desarrollarse en cooperación con los lugares recomendados.

Entre los propósitos del proyecto de investigación de los ecosistemas tropicales estaría el establecimiento y la obsevación contínua de parámetros claves del ambiente físico y biológico de cada lugar: el agua

Resumen y Recomendaciones 23

y el ciclo nutriente; la energética del ecosistema, los ritmos estacionales en la producción de hojas y frutas; la ecología fisiológica de las plantas; lo herbívoro; las redes alimenticias; y la dinámica de microambientes y de manchas. Anticipamos que los conocimientos recopilados en cada lugar afectarán aquellos recopilados en otros lugares y que el resultado aportará apreciaciones cuantiosas sobre el funcionamiento de los ecosistemas tropicales en general. Los estudios deben ayudar a establecer los factores incluidos en la creación de ecosistemas continuamente productivos en los trópicos e indicar las condiciones básicas en muchos sectores de la ecología y de la biología de la población.

En el campo de la investigación de suelos ya tenemos amplias correlaciones entre el suelo y la vegetación en los trópicos estacionales y en los trópicos húmedos en Asia y partes del Africa, aunque queda por hacer más para corroborar los modelos establecidos en Sur América. La interacción dinámica entre los bosques y suelos en diferentes climas y sobre distintos suelos es todavía poco conocida. Es necesario un estudio detallado del movimiento de nutrientes y los iones de microelementos en solución a través de las plantas, la fauna y del suelo. Las comparaciones entre los bosques primarios y varios tipos de tierra convertida proporcionarán una base esencial para el desarrollo de diversos usos sostenibles para las tierras. Dichos estudios se concentrarían, sin por eso llevarse a cabo exclusivamente, en los principales lugares descritos en este informe. En el campo de ecología fisiológica de plantas tropicales, recomendamos el establecimiento de un laboratorio comprensivo en Puerto Rico el cual se convertiría en un centro para la investigación y la preparación en este campo. La creación de dicho laboratorio parece esencial al establecimiento de un campo de investigación poco desarrollado en los trópicos. De esta investigación se espera una gran retribución en los científico y social.

ESTUDIOS DE LOS SISTEMAS ACUÁTICOS TROPICALES

Opinamos que por su importancia científica y económica los sistemas tropicales ameritan un estudio mucho más intensivo que el actual.

Los Ríos

Sugerimos iniciar esfuerzos inmediatos para apoyar y respaldar el estudio que se realiza actualmente sobre el Amazonas y el Orinoco y sus ramales principales. El Zaire también debe tener alta prioridad en el estudio. Los estudios de estos ríos deben incluir a la química del agua y el proceso de nutrientes, planctones, peces, la fauna del fondo, los vertebrados claves y las interacciones entre tierra y agua. En cuanto a los ríos más pequeños, debe prestarseles una atención prioritaria al Río Musi en Sumatra y el Río Purari en Papua Nueva Guinea y a otros ríos igualmente amenazados por cambios. Los estudios intensivos deben continuarse con una vigilancia a largo plazo de todos estos ríos.

Los Lagos

Destacamos los lagos de cuenca cerrada, los lagos que tienen mayor conjunto de especies endémicas y los lagos sujetos a cambios en el futuro inmediato. Al combinar estos criterios, llegamos a la conclusión que el estudio de los siguientes lagos es urgente: el Lago de Valencia y el Lago de Maracaibo en Venezuela, el Lago Malawi en Africa, el Lago Titicaca en Sur América y los lagos volcánicos en el sureste asiático insular deben estudiarse pronto.

Las Tierras Húmedas

Las tierras húmedas son de gran significado ecológico y son muy vulnerables a la destrucción. Le damos la más alta prioridad al estudio del Sudd, un pantano muy extenso de carácter único mantenido por el Nilo. Ya se han iniciado proyectos importantes de drenaje en este pantano. El Pantanal de Mato Grosso en Brazil, que en épocas anuales tiene mayormente tierras secas, es aún más grande y posiblemente sea de mayor importancia regional. Se enfrenta a un desarrollo activo. De prioridad menos urgente situamos los pantanos tropicales del Territorio Amapá en Brazil, los de Beni en Bolivia, y el Pantano Benguelu en Zambia, principalmente por una menor amenaza en el futuro inmediato (en los 5 años). De las tierras pantanosas a lo largo de ríos parecen estar amenazadas la *várzea* de la Cuenca del Amazonas y el delta y las aguas estancadas del Orinoco en lo inmediato; las aguas estancadas del Zaire y Xingu parecen menos amenazadas. Los extensos pantanos de musgo del sureste asiático necesitan atención por su extensión y por ser un fuente importante que se va explotando a un paso creciente.

La Precipitación

La química y cantidades de precipitación, que son importantes a las características de ambientes tanto terrestres como de agua fresca deben ser vigilados y estudiados, especialmente en el Amazonas, Orinoco y el drenaje del Mekong.

Resumen y Recomendaciones 25

VIGILANCIA DE LA CONVERSIÓN FORESTAL

Los esquemas nacionales para vigilar la tasa de conversión de los bosques húmedos tropicales y otros tipos de vegetación tropical deben fomentarse y cuando sea adecuado deben ser ayudados por organismos internacionales competentes. Ver NRC (1980).

PLAN QUINQUENAL

Durante el período de 1980–1985, pedimos que se tomen las siguientes acciones como asuntos de suma importancia para la atención de todos los paises del mundo:

- Por lo menos doblar, en dólares constantes, los fondos asignados la inventario biológico de los trópicos.
- Incrementar al menos por 50% el número de profesionales sistematistas dedicados a estudios de los organismos tropicales.
- Iniciar operaciones al menos en cuatro lugares de ecosistemas básicos ya mencionados, establecer un centro para el estudio de la ecología fisiológica de plantas tropicales, y completar la instalación de sus instalaciones básicas. Completar también los estudios básicos de los ciclos minerales y los inventarios de suelos y los biológicos de base en cada lugar de ecosistemas.
- Iniciar o aumentar los estudios comprensivos principales del Amazonas, Orinoco, y Purari con sus principales ramales; de los lagos de Valencia y Maracaibo; del Sudd, del Pantanal del Mato Grosso, el *várzea* de la Cuenca del Amazonas y del delta y de las aguas pantanosas del Orinoco. Los estudios quinquenales básicos que hemos resumido deben completarse en todos los casos antes de 1990, y los estudios de otros temas deben iniciarse en el período de 1986–1990.
- Asignar fondos para la vigilancia y la información internacional de las tasas de conversión de los tipos de vegetación tropical, especialmente los bosques húmedos tropicales.

2 The Problem

The lowland tropical forests of the world, which a century and a half ago comprised an area twice the size of Europe, have now been reduced to roughly half of their former extent. Many persons believe that these forests are being altered at a rate that will result in their disappearance as a natural ecosystem within our lifetimes (Iltis, 1972; Janzen, 1972, 1974a, 1977a; Richards, 1973; Whitmore, 1975; Myers, 1976, 1979a; Raven, 1976; NRC, 1980). Even though tropical moist forest may persist in western Brazilian Amazonia and Central Africa for another 40 or 50 years, in most other areas it will be reduced much sooner to scattered, degraded remnants on steep slopes, to severely flooded delta areas, and to a few parks and reserves.

Current human populations in these regions will have doubled by early in the next century, and the demand for food and other resources by these people is increasing much faster than their absolute numbers. Moreover, economic pressures operating on a world-wide scale engender activities that drastically alter vast areas of these forests. The very poor and malnourished people who constitute about a quarter of the world's population often have nowhere to turn for food except to their own native forests—but there are not yet, for the most part, technologies available by which they may produce food from these forest areas on anything but a short-term and then destructive basis. As a consequence, an area of tropical lowland forest about the size of Delaware is permanently converted to other forms each week, and an area about the size of Great Britain every year.

The Problem

Accompanying this destruction will be an extinction of plants and animals at a rate unprecedented in the history of the world. Since about two thirds of the species of most groups of organisms occur between the Tropic of Cancer and the Tropic of Capricorn, the rapid changes in tropical ecosystems that we are witnessing today could lead to an enormous and irrevocable loss of species and the associated opportunity to gain knowledge and to a permanent alteration in the course of evolution on a global scale through a drastic decline in genetic diversity. In addition, the destruction of these vast ecosystems, without the development of ways for replacing them with others equally productive, foredooms a large portion of the human race to misery and portends instability for the entire globe by the year 2000.

If, then, there is every reason to believe that more than half of the species of plants and animals that exist today will have become extinct by the year 2100, what we and our successors will have learned about them and the ecosystems in which they occur depends to a large extent on our ability to plan and act now. Many opportunities to increase scientific knowledge will be lost irretrievably, but many can be gained and the extinction rate reduced if we have the wisdom to lay a proper foundation and to act on it. A more widespread application of existing knowledge to management of forest, forage, and food crops, coupled with rapid expansion of fundamental biological information, appears to be the most promising approach to the solution of the awesome dilemma with which we are faced. It is our great collective tragedy that only a small effort is being devoted to alleviating this drastic situation, while a number of people roughly equal to the population of France or of the United Kingdom is being added to the tropics every year.

The theme of the Twelfth General Assembly of the International Union for the Conservation of Nature and Natural Resources in Zaire in September 1975 was conservation for development. At these meetings there was a strong call for incorporating biological principles into the development plans of tropical countries, as many students of the problem have long urged (recent examples include Dasmann *et al.,* 1973; Janzen, 1973a; Farnworth and Golley, 1974; Tinker, 1974; Goodland and Irwin, 1977; and Hawkes, 1979). The delegates called for international help in bringing this about and for legislation that would require companies active in the tropics to prepare and publish evaluations of the consequences of their activities.

Although the scope of the committee was limited to the study of biological processes, we consider it a matter of urgent priority that other committees and advisory groups be constituted to consider such matters as the world-wide interrelationships that largely govern the rate and kind of conversion of tropical forests. In addition, we recommend that urgent priority be given by social and biological scientists to the study of aboriginal populations throughout the tropics; with the destruction of the ecosystems, these people are disappearing rapidly. Additional information should be developed as soon as possible both because of its own intrinsic interest and because of the potential role of such knowledge in aiding the aboriginal peoples and others living in similar areas.

Building the human and institutional resources of the tropical countries offers the only hope for the future (see Farnworth and Golley, 1974; Prance, 1975; Ayensu, 1978). All scientists need to know more about the tropics in order to understand the limiting conditions for many of the biotic phenomena that interest them. But only the long-range study of these same phenomena by scientists resident in the countries concerned can yield a full understanding of the way in which tropical ecosystems function and thus lead to an improvement in the human condition. In addition, the effectiveness of a cooperative scientific activity in a tropical country should be evaluated not solely on the international significance of the scientific results in research but also on the extent to which that activity strengthens local capacities and infrastructures for research and training (diCastri and Hadley, in press).

As revealed by the indices of tropical research (Gómez-Pompa and Butanda, 1975, 1977) and the world census of scientists concerned with tropical ecology (Yantko and Golley, 1977), many resident scientists are pursuing fruitful research programs in tropical countries. Strong research and educational institutions, such as the Instituto Nacional de Pesquisas da Amazonia (INPA) in Brazil, the Centro Agronómico Tropical de Investigación y Enseñanza (CATIE) in Costa Rica, the Instituto Nacional de Investigaciones sobre Recursos Bióticos (INIREB) in Mexico, the Smithsonian Tropical Research Institute (STRI) in Panama, the Escuela Agrícola Panamericana (EAP) in Honduras, the Centro Internacional de Agricultura Tropical (CIAT) in Colombia, the International Institute of Tropical Agriculture (IITA) in Nigeria, the International Center for Research in Agroforestry (ICRAF) in Kenya, the Regional Center for Tropical Biology (BIOTROP) and the National Biological Institute (LBN) in Java, hold great promise for the future.

Scientists working in institutions such as the above have two important advantages that are denied co-workers who come from extratropical countries: (1) as residents, they can study the phenomena that concern them on a sustained basis and (2) they are often able to apply the results of their studies directly to human welfare in their own countries (Matthew, 1978).

The Problem

CONVERSION RATES OF TROPICAL MOIST FOREST

In 1975 the United Nations Food and Agriculture Organization (FAO) provided figures on conversion rates of tropical moist forest (Sommer, 1976). It is important to realize that these estimates refer to forest lands, not to residual primary forests, for which no estimates exist but which in Africa and Asia are only a small fraction of the remaining area. Similarly, they represent forest land at all altitudes; the lowland forests are the richest in species and also those that have been most severely altered. The 1975 analysis included the following information:

Potential area of tropical moist forest (world): 16 million km^2
Estimated area in 1975: 9.35 million km^2, or a reduction of 41.5% from potential limits
Conversion to date (1975), as proportion of original area:

	Potential Area (million km^2)	Reduction to 1975 (%)	Present Area (million km^2)
Latin America	8.03	37.0	5.06
Asia	3.87	43.7	1.87
Africa	3.62	51.9	1.75

The current conversion rate for tropical forests almost certainly exceeds 200,000 km^2 per year (Sommer, 1976; NRC, 1980). Myers (1979a) estimated that 245,000 km^2 of tropical moist forests are being converted to other purposes annually, and he subsequently confirmed this estimate on the basis of a more extensive and systematic study (NRC, 1980). For the three major regions of tropical forest, a World Bank report (1978) estimated annual conversion rates as follows: Latin America, 50,000 to 100,000 km^2; Africa, 20,000 km^2; Asia, 50,000 km^2. Even if these rates were constant, they would lead to a total destruction of all tropical forest world wide within 50 years. But the rates are not constant—they are accelerating rapidly.

In some countries, the rate of conversion is much higher than the average. In India, Sri Lanka, and Burma, nearly two thirds of the original area of tropical moist forest has already been converted to other purposes (Sommer, 1976), and the lowland primary forests are already reduced to a few isolated fragments. It is anticipated that the lowland forests of the Philippines and the Malay Peninsula and much of those of Indonesia, western Africa, Madagascar, Central America, and the West Indies will be converted from forests to other uses within the next 10 years. In contrast, sizable tracts of lowland forest in western

Brazilian Amazonia and Central Africa may persist into the first quarter of the twenty-first century, when they probably will be the only substantial areas of lowland tropical forest in existence.

Because inadequate information is available concerning the extent of forests of most tropical countries, it is urgent that a system be established for more-accurate monitoring of the rates of conversion of various kinds of tropical forest. In the course of this monitoring, it will be important to determine the various uses to which the forest is being converted, and, insofar as possible, the causes that are leading to its conversion. Information available on individual countries is summarized in NRC (1980).

LAND USE, POPULATION GROWTH, AND FOOD PRODUCTION

Patterns of land use and tenure in many tropical countries tend to inhibit overall planning and to encourage exploitative modes of development often linked with international trade (UNESCO, 1978). Thus, in 1975, 7% of the landowners in Latin America possessed 93% of the arable land; by contrast, in the United States, a comparable 7% of landowners had farms that included only 27% of the arable land (Eckholm, 1979a). The crops of these very large landowners in Latin America are often exported, with the frequent result that the bulk of the people must subsist on the productivity of a small fraction of the arable land. World needs for wood, beef, and other commodities are leading to the extensive destruction of broad tracts of forest to provide industrialized nations with these materials (Janzen, 1974a; Rappaport, 1978; Irwin, 1977; Myers, 1979a; NRC, 1980). Poor forestry practices are destroying the forests and reducing the potential yield. Increases in oil prices have made the production of fuel from such crops as manioc and sugar cane economically feasible and may ultimately lead to the clearing of large areas of forest in the tropics and elsewhere for this purpose.

As a result of these and other practices, and of rapid population growth, there exists throughout the tropics a large class of very poor people. As Eric Eckholm has put it, "More food production alone will not eliminate hunger; nor will more charity. Only secure access to decent land or jobs will give the dispossessed a chance to work their way out of extreme poverty and undernutrition" (Eckholm, 1979a). The World Bank has estimated that some 800 million people are living in what is called "absolute poverty"—that is, "a condition of life so characterized by malnutrition, illiteracy, disease, squalid surroundings, high infant mortality, and low life expectancy as to be beneath any reasonable definition of human decency" (McNamara, 1979). Given

The Problem

the situation as it is, the further clearing of forests in a manner that often seems uncontrolled is inevitable, whether the clearing is undertaken to produce exportable commodities for profit or a few crops for a poor peasant family.

The conversion of tropical forests is accelerated by the rapid growth of human populations in tropical countries. Against a world rate of population growth of 1.7% annually, the population of western Africa is growing at 3.0% per year, Southeast Asia at 2.2% per year, and tropical South America at 2.8% per year (Population Reference Bureau, 1979). Thus, for example, by the end of the century, the number of people in tropical South America will have increased from 193 million to about 332 million. About 90% of the total world population growth over the next 20 years will take place in the tropics. Contrasting with these figures are those for the industrialized world. The population of the United States is growing at 0.6% per year, that of the U.S.S.R. at 0.8%, and that of Japan at 0.9%; that of West Germany is dropping at 0.2% per year.

Of particular importance in evaluating the consequences of these projections are the world food and energy crises, which are intimately related. The relatively small amount of effort that is being expended world wide to improve the bases for tropical agriculture is unlikely to lead to the establishment of continuously productive systems in areas where tropical forest has been removed. Indeed, it is probable that productive agriculture will simply not be possible over large areas of the tropics no matter how much research is conducted. Many tropical nations lack energy reserves and are placing increasing demands on a diminishing supply of wood and dung for fuel (Openshaw, 1974; Eckholm, 1976, 1979b), thus further impoverishing the productive capacities of their soils. In the poorer nations of the world, as much as 86% of the wood consumed annually is burned as fuel (Arnold and Jongma, 1979).

Many tropical countries are simply assigning their primary forest tracts to foreign concerns without taking the opportunity to develop means for their sustained harvest (Meijer, 1973a, 1973b, 1975; Goodland and Irwin, 1975). By doing so, they are losing opportunities to understand how these complex ecosystems function and are therefore unable to gain much important economic and scientific knowledge.

At the First Conference on International Cooperation in Agroforestry, held at Nairobi, Kenya, in July 1979, it was estimated that at least 250 million people live as forest farmers in the tropical regions of the world (see also Myers, 1979a; NRC, 1980). As their populations increase and the forest disappears, they are becoming less able to practice the forms of shifting agriculture that were possible when the forests were more extensive and the people using them less numerous. These forest farmers use about a fifth of the entire biome (Sommer, 1976), and their activities constitute the largest factor in conversion of tropical moist forest (Myers, 1979a; NRC, 1980). When present in low numbers, forest farmers regulate their conversion of resources, in part by ritual or cosmological behavior, in such a fashion as to keep the system from running down (Rappaport, 1968; Reichel-Dolmatoff, 1975). When the people are too numerous, these traditional methods become ineffective. Moreover, the forest farmers are being joined by increasingly large numbers of peasants, who, because of the lack of land elsewhere, are moving into forests where they adopt a contemporary version of slash-and-burn agriculture and then, especially in Latin America, often move on to cattle ranching (Heckadon, 1979). This leaves less and less chance for forest regeneration. Trends such as this form a major element in FAO projections that by 1985 some 26 tropical countries, with an aggregate population of 365 million, will be unable to provide sufficient food for their people.

The rate at which the world's population is growing, 1.7% per year (Population Reference Bureau, 1979), means that between now and the end of the century food must be found for about 2 billion additional people. Since well over a billion people are undernourished now, out of a total world population of some 4.4 billion, substantial changes would have to be made in food distribution to feed the total population adequately. If there is continued growth in affluence world wide, and if a genuine effort is made to solve the problems of malnutrition, the demand for food in 1985 will be nearly 50% greater than in 1970. For the developing countries, 70-80% more food will be needed in 1985 than was consumed in 1970 (Ehrlich et al., 1977). The International Food Policy Research Institute (1977) has estimated tht the 1985-1986 annual deficit of cereal grains in developing countries will range between 100 million and 200 million metric tons, the amount depending on the rate of increase that can be attained between now and then. In spite of this, many tropical countries favor monoculture cash export crops that yield foreign-exchange credit, perceived as necessary for development.

Despite some optimistic forecasts, there is little evidence that the problem of world food supply will be solved by increased food production in developed countries and distribution of the surplus throughout the world (Schneider, 1976). The search for more-appropriate agricultural technologies for the tropics continues (Dahlberg, 1979), but there is little available arable land not already being cultivated. Indeed, 50,000-70,000 km^2 of cropland—an area the size of Belgium plus Holland, or of West Virginia—are being lost to erosion and related forces each year (Myers, 1979a).

As food becomes scarce, people in such countries as Bangladesh, India, and Haiti and in regions such as the Sahel die in increasing numbers and in relation to the amount of food available (Brown, 1976). With the present rate of world population growth and levels of life expectancy, the outlook is distinctly gloomy.

Because of these factors, including the operation of a world economic structure that favors the production of export crops in the tropics, no more than scattered remnants of undisturbed tropical lowland forest are likely to survive into the twenty-first century. Even at present minimum estimated rates of destruction—and the rates are being accelerated yearly—all tropical forest will be gone by the middle of the next century.

PRESERVATION OF THE FORESTS

Special arguments can be made for the importance of preserving areas of the tropics. For one thing, conservation is important because the number of species found there is much greater than in other regions (Terborgh, 1974). For most areas in the tropics, however, the facts about which species are actually found and which are not, or where else these species occur, are not known for the great majority of all plants and animals. Nonetheless, the preservation of adequate tracts of tropical lowland forest is one obvious way in which the human race has an opportunity to retain some of its options for the future (Myers, 1979a). Despite this, the conservation movement in the tropics is still in its infancy, with less than 1% of the tropical biome preserved in parks and reserves at present (Bernardi, 1974; Myers, 1976, 1979a; Whitmore, 1976; Prance and Elias, 1977). In line with the objectives of Project 8 of UNESCO's Man and Biosphere Program, the establishment of an international network of biosphere reserves that would extend through the tropics is a matter of high priority. These areas can serve as base lines or standards by which the effect of human activities outside the reserves can be judged.

If such reserves are to be planned effectively, the concepts of modern population biology must be brought to bear on their design and management (see Diamond, 1972, 1975, 1976; Terborgh, 1974; May, 1975a; Simberloff and Abele, 1976; Myers, 1979b; Soulé and Wilcox, 1980). Political and economic constraints, however, often dictate that the sizes and locations of ecological reserves are not what would be optimal from an ecological viewpoint if ecological considerations had played a more prominent role in their establishment. The main role of ecologists, therefore, will often be to provide insights into the proper management of established reserves so as to preserve the maximum biological diversity as well as the key organisms and associations of organisms for which the reserve may have been established originally. Such management schemes must be founded on an understanding of the interactions among the component species (Janzen, 1974a). The role of the ecologist begins, rather than ends, when reserves are established. Certainly the kinds of research outlined later in this report will enhance our ability to make wiser decisions about the management of such reserves in the future.

As a consequence of the rapid destruction of tropical forests world wide and the mounting economic pressures on the remaining forests, most of the parks and reserves that can ever be established in the tropics must be established during the next 10 years. Only a limited amount of information is available about how best to manage and design these reserves (Wetterberg et al., 1978), and it is therefore important to accelerate research into this matter. One approach, for example, has been to try to set aside areas where the forest is thought to have survived during the dry periods of the Pleistocene, but there are distinct shortcomings to exclusive concentration on this approach (Lovejoy, 1979a), especially in view of the theoretical nature of the postulated refugia.

Since much of the rain forest that survives into the twenty-first century will be patches of formerly continuous forest, a study of the dynamics of such patches is of importance to research and management (Lovejoy and Rankin, 1979). It is known that forest fragments lose species after isolation (Terborgh, 1974; May, 1975a; Diamond, 1976). In this context, any kind of ecosystem seems to have a minimum critical size; loss of species is more of a problem below this size than above it (Lovejoy and Oren, in press).

The relative vulnerability of different species and ecosystem components to different kinds of intervention must be determined if effective conservation practices are to be implemented. In addition, the potential of various kinds of derived ecosystems as homes for particular species must be investigated in depth, and management plans must be developed accordingly (Soulé and Wilcox, 1980).

Natural ecosystems help to protect watersheds, recharge the water table, maintain patterns of rainfall (Lettau et al., 1979; Salati et al., 1979), and improve air quality; in short, they help to maintain environmental quality. For these reasons some countries—Brazil is a notable example—are giving serious consideration to placing severe restric-

The Problem

tions on further deforestation. The economic pressures, however, are enormous and often give rise to intense political pressures. If more than half of the precipitation in the Amazon region is forest generated, as was recently calculated (Villa Nova et al., 1976; Salati et al., 1978), then such protection is clearly essential for the future environmental health of the region. The careful monitoring of precipitation in relation to forests is clearly of great importance.

Forest plantation projects are of interest for their potential reorientation of the production of tropical wood requirements toward small areas of concentration, in lieu of current removal of a few trees per unit of area from extensive primary forests. If timber production and processing could be concentrated and combined with integrated food production, it appears that there would be an overall reduction of stress on primary forests. The maintenance of trees, even in plantations, appears to offer far less serious environmental consequences than any scheme by which they are removed, but the consequences on the soil biota of introducing alien trees, often ectomycorrhizal, also need to be considered.

It is a simple matter to assert the importance of setting aside reserves and parks in the tropics for their aesthetic value, as vital resources from which new scientific information can be derived, and as a basis for guiding sound development for human welfare (Myers, 1979a). If the development of such reserves is to be successful, however, the industrialized countries of the world must help to pay for their establishment and preservation (Nicholls, 1973; Goldsmith, 1980). The economic implications are apparently well beyond the expectations of the industrialized world (Myers, 1979a).

SCIENTIFIC RATIONALE FOR LEARNING MORE ABOUT THE TROPICS

The scientific reasons for accelerating research in the tropics are numerous. In temperate regions, perhaps a million species of plants and animals have been named, but in the tropics the comparable figure is only about 500,000. Investigations of groups of organisms that are relatively poorly known in temperate regions, such as mites and nematodes, suggest that several hundred thousand more species are yet to be discovered outside the tropics, or a total of perhaps 1.5 million species of organisms in temperate regions. Since in most well-known groups of organisms there are about twice as many species in the tropics as in temperate regions, one may surmise that there are probably at least 4.5 million kinds of organisms in the world, of which at least 3 million occur in the tropics. At least three quarters of these tropical species still await discovery and naming, and the actual numbers may well be significantly higher.

Our ignorance of tropical organisms at the most fundamental level—cataloging and naming—is illustrated by the vast discrepancies in estimates of numbers of species in the tropics as well as by the recent discoveries of previously unknown kinds of conspicuous higher animals, such as birds and mammals (see Casey and Jacobi, 1974; Wetzel et al., 1975; Fitzpatrick et al., 1977; O'Neill and Graves, 1977). It is essential that broadly based survey projects, such as the *Atlas of Neotropical Lepidoptera* (Smithsonian Institution), *Flora Malesiana, Flora Neotropica,* and *Projeto Flora Amazônica,* be supported adequately and accelerated in pace. The relevance of such projects to those resident in the countries concerned ought to be reexamined continually, and priority given to those in which it can clearly be demonstrated that increased funding would lead to their more rapid completion. All work in tropical biology is a race against time, and studies that can be used effectively in the field in the near future deserve to be emphasized.

Local accounts of the floras and faunas of particular areas likewise provide invaluable and original insight, illuminating the whole pattern of tropical diversity, and such studies should be encouraged. Outstanding recent examples are Duellman's (1978) detailed analysis of the herpetofauna of a limited area in Equadorian Amazonia and Croat's (1979) comprehensive account of the vascular plants of Barro Colorado Island in Panama. In selecting the sites for detailed studies of this sort, and for local floristic and faunistic accounts, emphasis should be placed on areas that are being altered most rapidly and on those that contain the greatest concentrations of primitive or otherwise interesting taxa. Examples are Madagascar, Hawaii, Borneo, Papua New Guinea, and portions of coastal Brazil (in the states of Bahia and a small part of Espírito Santo). In addition, areas in which speciation seems to have been especially prolific, such as the lakes of the East African Rift Valley, should receive special attention.

Especially critical is the shortage of trained systematists, both in tropical and temperate regions, who are competent to identify and classify animals and plants. The situation is especially acute for many groups of insects, some of which are of great economic importance. If the terrestrial and freshwater organisms of the tropics are to be described and studied in detail, immediate efforts must be made to recruit and train large numbers of systematists and to provide funds for their continuing support. The efforts of systematists are indispensable for advances in all fields of tropical biology, including ecology, as well as such applied fields as agriculture and conservation.

If such a large proportion of the plants and animals of the tropics is undescribed, it is no surprise that we know virtually nothing about any feature of biological interest for the vast majority even of those that have been described. In the higher plants of the tropics, for example, little information is available about chromosome number and chromosome morphology (Bawa, 1973; Raven, 1975). Some intensive studies of particular groups must be made if we are even to have an index to the ecological and evolutionary patterns displayed by the remainder. In addition, appropriate groups of tropical organisms should be selected for detailed studies of the mechanisms of adaptation, speciation, and microevolution they exhibit. It is of considerable theoretical importance to determine how closely the patterns that are discovered correspond with those that have been identified in analogous groups in temperate regions.

At the level of ecosystem processes, our ignorance of tropical systems is profound. One example of the unique nutrient relationships of tropical forests is the pathway of phosphorus and nitrogen exchange in them. It was proposed by Went and Stark (1968a, 1968b), and has now been confirmed by Stark and Jordan (1978) and by Herrera *et al.* (1978a), that the roots of some tropical rain-forest trees, growing on the least-fertile soils, can directly recapture phosphorus and nitrogen from leaf litter through mycorrhizae. There no doubt exists a wide array of mechanisms by which tropical plants acquire and distribute nutrients in an environment where stores of soil nutrients are low because of rapid leaching. Discovery of these mechanisms could have far-reaching implications for the management of the composition of tropical forests on stressed sites.

Similarly, the nature and role of secondary vegetation and succession in the tropics have been explored only from the standpoint of wood production, not from either a purely ecological or evolutionary standpoint (Gómez-Pompa, 1972; Gómez-Pompa *et al.*, 1976; Hartshorn, 1978; Ng, 1978; Whitmore, 1978). A better understanding of the genecology of tropical plants is important for both theoretical and practical reasons, since there is no *a priori* reason to assume that they resemble better-known temperate plant groups in these respects (Ashton, 1969, 1970, 1976a; Bawa, 1976). We know very little about the demography of plants in general and almost nothing about tropical ones (see Sarukhán, 1978, 1979). The study of the architecture of trees, a science of the 1970's, has begun to analyze in detail the rich diversity among tropical trees that is now seen to play a critical role in the structure of tropical ecosystems and to affect profoundly the flow of energy in them (Rollet, 1974; Hallé *et al.*, 1978). A proper understanding of the operation of pollination systems in tropical forests might hold an important key to understanding the evolution of floral diversity among angiosperms (Schnell, 1970; Bawa, 1974; Tomlinson, 1974; Baker, 1975; Bawa and Opler, 1975; Arroyo, 1976; Burley and Styles, 1976). The development of answers to questions such as these must proceed rapidly because the vegetation of many areas inhabited by large numbers of primitive angiosperms, such as New Caledonia and Madagascar, is rapidly being destroyed.

We also do not know whether tropical forests are stable because of their great complexity or whether they are very fragile and easily disrupted (Farnworth and Golley, 1974; Windsor, 1974, 1975, 1976; May, 1975b, 1979; Connell, 1978). The proper understanding of this issue has important theoretical and practical implications relating to the origin, functioning, and maintenance of global trends in biological diversity and the consequences of agricultural and forestry practices.

The conversion of areas now occupied by tropical forests to other uses must have already caused a permanent alteration in the course of evolution world wide (Richards, 1952). It will also diminish our ability to answer many significant biological questions that are evident to us now, let alone those that might be asked in the future. The increasing utilization of tropical examples in such important and growing fields as sociobiology has been analyzed by Janzen (1977b). In addition, as Richards (1973) and Robinson (1978) have stressed, we simply need to know more about the excellent examples of adaptation that are found in the tropics, and thus add to the store of knowledge about the planet on which we live. For example, ant–acacia interactions (Janzen, 1967), army ant–bird interactions (Willis and Oniki, 1978), evolution in Hawaiian *Drosophila* (Carson and Kaneshiro, 1976), and periodic mass flowering in bamboos (Janzen, 1976; Soderstrom, 1979) are tropical phenomena, each of substantial intrinsic interest and importance. How can a bamboo count 120 years? In view of all these considerations, it is not surprising that a group of systematic and evolutionary biologists convened in St. Louis, Missouri, in 1974 concluded that research in the tropics merited the highest priority (Anonymous, 1974a).

Opportunities for study of tropical freshwater environments are being lost as rapidly as, or possibly even more rapidly than, those for the study of terrestrial ecosystems. In this, as in all aspects of tropical biology, it is important to study first those ecological systems and biological phenomena that are confronted with imminent drastic change. It is also critical, in view of the short time available, to select systems and phenomena for study that are likely to have the greatest scientific significance and the most obvious relationship to the critical problems of human welfare in the tropics.

The Problem

The study of river systems, threatened on a world-wide scale by watershed deforestation, impoundment for energy production and flood control, and industrial and urban pollution, is critical to provide base-line data against which to measure future change, and, it might be hoped, influence the decision-making process responsible for the change. The largest rivers and largest tributaries are unique, few in number, and highly subject to change in the immediate future. We know relatively little about the organic and inorganic materials in tropical rivers, and knowledge about them is important to understanding watershed functioning (Sioli, 1975). In particular, the composition and dynamics of the river plankton have been studied very little (Talling, 1976). For the next decade or two these communities may still be studied in a relatively unaltered state in very large river systems. The organization of communities of fishes and invertebrates in tropical rivers is poorly understood (Lowe-McConnell, 1975), and these populations are facing radical change.

Similar considerations pertain to tropical lakes and swamps. Studies of tropical lakes will be highly useful in unraveling the role of the physical and chemical environment in controlling their biota. Some striking contrasts between tropical and temperate lacustrine systems are already suspected (Lewis, 1979). The interface between freshwater and terrestrial ecosystems is of special importance. The study of this interface, as represented by freshwater wetlands, should be accorded high priority because of the great vulnerability of such areas to change.

The quantity and quality of precipitation deserve special attention in the tropics. Even the amounts of precipitation in some extensive areas are not well known. The relationship between climate and forest cover urgently requires intensive study. Precipitation chemistry, although of great interest, is little studied. Priority should be given to areas in which there is great likelihood of major changes in forest cover. It is likely that both the amount and the annual distribution of precipitation in the tropics are closely correlated with forest cover and with vegetation in general.

Another major gap in our scientific information concerns the functioning and adaptation of human populations in the tropics. Aboriginal populations have long existed in tropical forests and possess considerable knowledge about them. Human diversity as well as biological diversity is being reduced world wide, and the rich variety of conceptual and agricultural expression that characterizes the thousands of distinct groups of people living in the tropics is of great theoretical interest; studies would probably contribute significantly to our understanding of the development of human thought processes.

If we are to learn more about human attitudes in aboriginal groups in the tropics, we must assign urgent priority to this area because traditional human populations are among the first elements to be disrupted. The next 10 years will be of crucial importance for our ability to learn anything about most of these groups of people, including the ways in which they support themselves in the forest. Such information is of importance both for their future welfare and for ours. Although we have considered such studies to lie beyond the scope of our committee, we urge that they be emphasized and that an overall scheme for carrying them out efficiently and on a wide scale be developed as a matter of high priority.

SOCIETAL RATIONALE FOR LEARNING MORE ABOUT THE TROPICS

The anticipated destruction of virtually all tropical vegetation during a quarter of a century when the human population of the tropics will double provides ample social and economic reasons for gathering more facts upon which any improvement of human welfare in the tropics may be based. Moreover, the ramifications of an ecological change of this magnitude are so far reaching that no one on earth will escape them. We need to use wisely and expeditiously every opportunity that we have to study the functioning of tropical ecosystems and to apply that knowledge directly to the real and growing problems that confront human populations in the warmer parts of the globe.

Although soils in some limited areas of lowland tropical forest are relatively fertile, most are extremely infertile. In addition, temperature and rainfall reduce the effectiveness of temperate-zone agricultural methods in the tropics. For example, it has been estimated that less than 10% of the land in the Amazon Basin can support sustained agriculture, even when fertilized, because of the properties of the soils (NRC, 1972; Irion, 1978). Consequently, agriculture ought to be concentrated in the most suitable areas. Budowski (1974, 1977) has advocated the establishment of forest plantations on already degraded sites to deflect development and exploitation away from the remaining primary forest (see also UNESCO, 1978).

Janzen (1973a) has provided an illuminating discussion of some of the factors involved in the development of sustained-yield tropical agriculture and forest harvesting and has outlined some of the kinds of knowledge required before new systems will be ready for adoption. The overall question is how to develop high sustained yields and high gross productivity in areas occupied by forests in the tropics. Studies for the production of timber from already regenerated forests have been developed in Malaya (Wyatt-Smith, 1963; Meijer, 1971), Uganda (Dawkins,

The Problem

1958), and Nigeria (Lowe, 1975), but techniques for generating subsequent crops naturally are still untested. If primary tropical lowland forest is cut, it cannot be replaced and usually will not regenerate as it was, even if small reserves are established (Gómez-Pompa *et al.*, 1972). Nevertheless, the potential of reserves of limited size in the reestablishment of native forests requires much further study, as do the factors controlling succession in tropical forests (see Gómez-Pompa *et al.*, 1976).

The production of timber on infertile tropical soils preserves forest cover and offers watershed protection benefits (Budowski, 1974). Natural regeneration can be manipulated for sustained timber yield, and the structure of the forest can be altered by selective removals to enhance such yield in the early regeneration (Fox, 1976). Simplified mixes of native species of relatively uniform age, however, are easier to manage (Wyatt-Smith, 1963).

Because forest planting has sometimes led to yields five times as high as those of managed native forests in many countries, more-rapid growth of plantations has all but terminated timber production investments in native forests (Lowe, 1975; Auchter, 1978). Such methods are still practiced in the dipterocarp forests of the Far East, however (Hutchinson, 1979). Since secondary forests support numerous native species of plants and animals, their management tends to preserve selected gene resources. A majority of all species, however, can exist only in primary forest, and they regenerate very poorly (Gómez-Pompa *et al.*, 1972).

One of the most important reasons for learning more about the tropical forests concerns the widespread climatic effects that could accompany their destruction (Farnworth and Golley, 1974; Schneider, 1976; Villa Nova *et al.*, 1976; Friedman, 1977; Lettau *et al.*, 1979; Myers, 1979a; Salati *et al.*, 1979). It has been suggested that if rainfall is decreased in the equatorial zone, an increase in precipitation between 5 and 25 degrees north and south and a decrease between 40 and 85 degrees north might follow, with possibly serious consequences for the main grain-growing belts of North America and Eurasia (Poore, 1976). The actual effects will, of course, depend on the nature of the vegetation that replaces the primary forest.

In addition to the effects on precipitation, a number of observers believe that widespread conversion of tropical forests to pastures and fields could contribute significantly to a worldwide buildup of carbon dioxide that could well affect climatic patterns throughout the world (Hutchinson, 1954; Potter, 1975; Potter *et al.*, 1975; Molion, 1976; Stewart, 1976; NRC, 1977; Stumm, 1977; Woodwell *et al.*, 1978;

Broecker *et al.*, 1979; Choudhury and Kukla, 1979; Crutzen *et al.*, 1979; World Meteorological Organization, 1979; Wigley *et al.*, 1980). The extent to which the clearing of tropical forests contributes to the amount of CO_2 in the atmosphere is a matter of current debate, but as widespread burning occurs in many tropical forests, forests may become a significant source of atmospheric carbon dioxide (Bolin, 1977; Woodwell *et al.*, 1978; World Meteorological Organization, 1979), with major effects on world climate. Twice as much carbon dioxide in the atmosphere could well cause an average global increase in temperature of 2–3°C, a greater change than has occurred during the past 10,000 years. One result could be warmer and drier weather in the grain-growing areas in North America, with all that entails for the United States' capacity to feed itself and many other countries. Indeed, it is believed that a temperature increase of only 1°C could decrease U.S. corn production by as much as 11% (Bryson and Murray, 1977; d'Arge, 1979; World Meteorological Organization, 1979).

Many of the most important, and most intractable, problems of developing appropriate modes of utilization of tropical forest in ways that can be sustained have to do with interactions between organisms. In the tropics, there is, for the most part, no counterpart to the detailed natural history studies that underpin most temperate-zone ecology. The intensity of predator–prey and herbivore interactions in the tropics, for example, and their incredible diversity and frequent specialization, have led scientists to believe that the direct application of temperate-zone pest-control methods to the tropics is questionable (Gray, 1974; Janzen, 1973a; Lamb, 1974; UNESCO, 1978). There is also concern that increases in the roughly 200,000 metric tons of pesticides now applied annually throughout the world may cause the destruction of whole ecosystems.

It is also imperative to accelerate the pace of biological inventory in the tropics for the potential identification of new kinds of useful and injurious plants and animals (Harney, 1979; Myers, 1979a). It would be of special importance to inventory members of those groups of organisms known to be important as herbivores or plant pathogens and to find those associated with cultivated plants and their relatives. To cite just one example, the brown rice leafhopper (*Nilaparvata lugens*), now ravaging about a third of the rice crops in tropical and subtropical Asia, is virtually unknown biologically. We know essentially nothing about its feeding and behavioral habits, migration, breeding, or even separation from related species (Yanchinski, 1978). Its explosive population growth appears to be related to changes in rice production in Asia associated with the introduction of new rice varieties and new cultural methods. The new rice varieties are no longer resistant to the pest, as

The Problem

are the old ones, which had been selected by farmers over many years. As a result of the damage done by this leafhopper, coupled in some areas with flooding, the government of Vietnam called for international famine relief for some 1.7 million people in late 1978. There are no more than a handful of systematists world wide who are competent simply to identify individual species of leafhoppers, of which there are tens of thousands, and which inflict billions of dollars' worth of damage annually.

The explosive spread of the brown rice leafhopper is but one example of the way in which the clearing of tropical forests and the reduction of overall diversity has led to epidemics of various diseases; the history of trypanosomiasis in Africa provides a classical example of this sort (Payne, 1980).

Whatever their potential economic role, most of the plant and animal species of tropical moist forests are highly vulnerable to extinction. Of all the biome's species, it is tentatively estimated that around 70–75% are arthropods, the great majority of them insects. Many of these species have highly specialized ecological requirements, many exist at low densities, and many are confined to local areas. These three attributes alone mean that these species are highly vulnerable to elimination when a tract of forest is subjected to the more disruptive forms of conversion, even if it is not entirely cleared. Some have the ability, like the brown rice leafhopper, to become widespread pests, but most will simply become extinct.

In view of the conclusions reached in recent surveys of conversion rates of tropical forests (Myers, 1979a; NRC, 1980), it is not unrealistic to suppose that, within the next two decades, as many as a million species of plants and animals could disappear in these forests, with another million likely to vanish during the twenty-first century. Extinction of a species is an irreversible loss of a unique resource. Thus it differs from more-common types of environmental impoverishment, such as pollution. When water bodies are fouled or the atmosphere is contaminated, we can, if we change our minds about the process, clean up the pollution. But species extinction is final—and the consequence falls not only on present society but also on all generations to come. Extinction of a species may have a direct effect on our everyday living. Throughout the world, people increasingly consume foods, take medicines, and use industrial materials that are derived from animal and plant species in tropical forests. These species' stocks can be reckoned among society's most valuable raw materials. Any reduction in the diversity of our resource base restricts our capacity to respond to new problems and opportunities.

Among plants, for example, two of the most important new crops that became of world importance during the past hundred years—rubber and the oil palm—have come from tropical forests, and many more potential crops are undetected or underutilized. A start in putting some of them to use was recently made (NRC, 1975, 1979).

The preservation of genetic diversity, both in plants and animals of known economic importance and in those for which relevance to human welfare is yet to be established, is properly a matter of international concern (Ehrenfeld, 1972, 1976; Iltis, 1972; Frankel and Hawkes, 1975; Pye, 1976; Frankel, 1977; Lovejoy, 1979b). The consequences of the loss of potentially important crops and of the loss of genetic diversity in established ones and their wild and weedy relatives may be serious (Heslop-Harrison, 1973, 1974; Hawkes, 1977a, 1977b, 1979; Whitmore, 1980). For example, the immense virus-resistance potential of the recently discovered perennial wild maize (teosinte) from Jalisco, Mexico, might alone have far-reaching agricultural implications (H. Iltis, personal communication, University of Wisconsin, 1980). This economically significant plant is now known from an area of less than 2 ha!

Considering the fact that less than a tenth of the estimated 150,000 species of tropical angiosperms have been screened for even a single class of chemical compounds, it is reasonable to expect that the remaining plants will ultimately provide numerous new compounds that will prove useful in medicine and for other applications (Whitmore, 1975; Myers, 1976, 1979a). Many more kinds of plants will be found, when they have been surveyed, to be sources of products useful for industry, such as gums, latex, resins, dyes, waxes, oils, and sweeteners, not to mention new sources of energy (Calvin, 1976; Marzola and Bartholomew, 1979; Myers, 1979a). The uses of native plants by indigenous peoples will doubtless continue to be useful in indicating potentially valuable ones (Haantjens, 1975; Hanlon, 1979). For example, more than 260 kinds of plants are known to be used in South America alone to control fertility (Moreno and Schvartzman, 1975). Even that kind of information, however, is largely ungathered and could disappear with aboriginal cultures.

Finally, tourism both by residents of the countries concerned and by foreign visitors constitutes an additional economic argument for the preservation of tropical forests and for moderation in the development of the remainder (Ehrenfeld, 1972, 1976; Meijer, 1973; Whitmore, 1975; Budowski, 1976; Myers, 1976; Western and Henry, 1979). Tourism provides great opportunities for education, leads to an enhanced appreciation of nature, and can be a good source of foreign credit. Scientific research could go hand in hand with the development of selected areas

The Problem

for tourism and could be used to provide instructive and interesting examples of evolutionary processes. The increased utilization of local examples in primary and secondary schools and the development of texts based on them would help to increase the appreciation of natural forests everywhere. Unfortunately, most available texts offer few examples from the tropics.

In recent years, an enhanced appreciation of what Ehrenfeld (1976) has called the "natural art value" of tropical forests, and of nature in general, has begun to develop. Many other kinds of values have been assigned to tropical forests, and they are well summarized by Poore (1978); see also Budowski (1976). More and more people believe that other species have a right to exist with us on the planet; for example, the rapid disappearance of primates is widely regarded as undesirable (Thorington and Heltne, 1976; Kavanaugh, 1979). The recent controversy about the right of people to decimate a dwindling population of some 50,000 chimpanzees to produce a relatively innocuous vaccine to treat a disease that is rarely fatal, hepatitis B, illustrates the point very well (Wade, 1978).

3 Inventory of Tropical Organisms

A comprehensive understanding of ecosystems must ultimately depend on a basic knowledge of the organisms that make up these systems. It is necessary that the component species be recognized as distinct units if knowledge about them is to be acquired and applied in a meaningful way. It has been estimated that there may be no fewer than 3 million kinds of organisms in the tropics, of which only about 500,000 have been described by taxonomists since Linnaeus established the current system of naming organisms. When we consider the extremely limited amount of information available concerning such huge groups as most insects, mites (Wharton, 1964; Stanton, 1979), nematodes, and fungi, we realize that the actual number may be much greater than 3 million. In any case, it is clear that the task of cataloging this diversity is very large. The rapidity with which tropical vegetation is now being destroyed suggests that high priority in the immediate future ought to be accorded to collecting and identifying species in relation to other groups, rather than to exhaustive taxonomic study. Ways should certainly be found, however, to apply knowledge concerning these organisms to practical advantage as directly as possible. Basic systematic knowledge is of fundamental importance for all purposes, and the initiation of a crash program for specimen acquisition and rough sorting appears fully justified.

Inventory of Tropical Organisms

NUMBER OF SYSTEMATISTS

Unfortunately, there are few systematists who are qualified to participate in such inventory work. It is estimated that about 250 people in the United States are capable of making authoritative identifications of at least one group of insects (Sailer, 1969; Knutson, 1978), and there are probably not many more than twice that number in the world. By analogy, and considering the relative popularity of different groups of organisms among systematists, it may be estimated that there are not many more than 4,000 such scientists in the world at present. Many of these systematists work exclusively or primarily with temperate-zone organisms. Probably no more than 1,500 trained professional systematists are competent to deal with any of the roughly 3 million kinds of tropical organisms, and their actual number may be declining because of decreased professional opportunities, reduced funding for research, and assignment of higher priority to other disciplines.

These estimates indicate that a high priority ought to be set on training and support for much larger numbers of systematists oriented toward tropical organisms. At least a five-fold increase in the number of systematists is necessary to deal with a significant proportion of the estimated diversity while it is still available for study. Governments would be well advised to allocate resources in an effort to achieve this objective (ESRC, 1977).

The shortage of systematists can be highlighted clearly by reference to particular groups of organisms. We still have a great deal to learn even about the best-known groups of organisms, and for most the situation is much worse. Some fungi, for example, cause damage costing billions of dollars annually; others are beneficial, as in antibiotic production or in the maintenance of fertile soils. Despite this, there is not a single area in the tropics for which the fungi are even relatively well known, and it is impossible to prepare regional accounts for any but a very few groups on the basis of collections that are available. How many unknown fungi in the tropics are potential pests similar to the southern corn leaf blight (*Helminthosporium maydis*), which caused over a billion-dollar loss in the U.S. maize crop in 1970 alone (Tatum, 1971)? Nematodes are plant pathogens of the highest importance, inflicting losses comparable with those caused by fungi (ESRC, 1977; ABRC, 1979); yet the nematodes of Central and South America, except in agricultural situations, are virtually unknown. Native faunas of nematodes world wide are rapidly being displaced by introduced forms, and many large areas are no longer suitable for a determination of the original situation. Freshwater mollusks are well known as intermediate hosts for many of the more important tropical diseases; yet for perhaps a third of the named species of mollusks from South America, we have only the original collection and the name of a country or province (A. Solem, Field Museum of Natural History, Chicago, personal communication, 1979).

Virtually all groups of tropical insects are poorly known. A number of insect and other arthropod taxa, such as ants, termites, and bees, are of obvious ecological importance; others, including soil and parasitic mites, mosquitoes, and sandflies, are well-known vectors of disease. Others that might receive special attention include aquatic insects as indicator organisms, parasitic wasps as biological control agents, and plant pests, such as leafhoppers, aphids, tephritid flies, and thrips. Of parasitic wasps, for example, there are probably at least several hundred thousand species in the tropics, each, so far as we know, highly specific in parasitizing a particular kind of insect. Finally, increased emphasis on the study of certain groups that are relatively well known, such as butterflies and many beetles, might be justified because of the probability that their study would yield results of scientific significance more rapidly than the study of other groups that are initially less well known.

Even among the vertebrates, the best-known group of organisms, many species remain undiscovered in the tropics. It has been estimated that as many as 40% of the freshwater fishes of South America have not yet been classified scientifically (Böhlke *et al.*, 1978), and the fishes of tropical Asia are poorly known also. As for birds and mammals, there are apparently relatively few undescribed species—a fair number have been discovered in recent years—but we know nothing except a name and a few localities for many of them.

Of the flowering plants of the world, basic to the structure of nearly all tropical ecosystems, about 155,000 out of 240,000 species are tropical, and about 90,000 are in Latin America (Prance, 1977). There are probably about 15,000 species of flowering plants in Latin America that have not been named. Others have been named more than once, and much revisionary work needs to be done. In addition, many other species have been collected only once or a few times, and it is virtually impossible to assess their status. For certain areas, such as the Chocó in western Colombia, we probably know only about a quarter of the species that exist. There could well be as many as a thousand unidentified species of endemic flowering plants in this area alone (A. H. Gentry, Missouri Botanical Garden, personal communication, 1979), and it is being deforested as rapidly as most of the other areas in South

America. Even for purposes such as establishing "ground truth" for aerial photographs, it is essential that botanical taxonomic research be intensified (Nielsen and Aldred, 1976).

METHODS AND CRITERIA FOR SAMPLING

For many kinds of tropical organisms, the collections that are made within our lifetimes are likely to be the only samples available for study in the future. The preservation of such samples, therefore, particularly if they are preserved in unusual ways (birds preserved in fluid, frozen tissue samples, and so forth), may be viewed as important simply for the sake of preserving for posterity some samples of the diversity of the earth as we find it now.

The basic sciences of systematic and evolutionary biology merge imperceptibly into service fields of the greatest importance for practical environmental work in pollution control, resource use, agriculture, and public health (NRC, 1970; Anonymous, 1974a). It is, however, patently impossible, given the present organizational systems, for information to be made available efficiently for such purposes. For a basic improvement in the situation, it would be necessary to develop sorting centers, such as those now in operation for marine organisms and for insects at the Smithsonian Institution, to process quickly and efficiently the material that is obtained and put into the hands of appropriate systematists (Lachner *et al.*, 1976). It would be necessary to augment subsidies for the operation of the large museum centers of international importance so that they could cope with the flow of specimens that would accompany such an accelerated effort. In addition, it is important to provide a higher level of support to museums and other centers in the tropical countries themselves, especially to those of the greatest national and international significance. The museums that house systematic collections have a vital role to play in collecting, storing, studying, and distributing the large amounts of biological material that will be accumulated. The training of indigenous systematic biologists is therefore of the utmost significance to this task.

Acceleration of biological research in the tropics would require electronic data processing for all aspects of systematic biology so that the facts, once ascertained, could be retrieved efficiently. Without it, we can hardly expect to learn much about the upward of 2.5 million undescribed species of organisms that occur in the tropics while there is still time. Data processing would open the way to development of new crops, to the screening of samples for potentially useful natural products and potentially harmful pests, to a better understanding of the functioning of natural ecosystems, and to the wise management, including conservation and preservation, of natural resources.

In selecting ways to approach this vast problem as effectively as possible, we must consider many interlocking criteria. Sampling of a group of organisms over a very wide area may often yield insight into its evolutionary pattern, but it seems that much can be said for intensive studies of as many groups of organisms as possible at particular sites. Certainly the biota of areas selected for intensive ecosystem studies should be sampled quickly for as many groups of organisms as possible.

In certain other areas—possibly, for example, at selected sites in the Amazon Basin—there might be a concentration of biological inventory. In this connection, the ecological stations set up by the Special Environmental Secretariat of Brazil's Ministry of the Interior might serve as a comprehensive network for biological inventory. Extensive efforts of this sort have long been carried out in Panama and elsewhere by the Gorgas Memorial Laboratory and by the Smithsonian Tropical Research Institute, by various branches of the French Office de la Recherche Scientifique et Technique Outre-Mer, and by other agencies. These efforts should be sustained and, when possible, amplified.

An example of an organization that appears to have special significance for this purpose is the Wau Ecology Institute, first established as a field station of the Bishop Museum, Honolulu. Located as it is in the center of Papua New Guinea in a rich midmontane environment with access to a great range of altitudes, it has already been used as a base for extensive biological inventory and could well serve as a locus for much more extensive efforts by the international biological community in studying the rich biota of New Guinea. At places like Wau, many groups could be dealt with simultaneously, with consequent gains in knowledge of the interactions among the organisms. If certain preselected sites were to be studied in this kind of detail, the records could be computerized and correlated with records of weather, soils, and other important ecological parameters. It stands to reason that the samples would facilitate any later effort to build up a comprehensive understanding of the structure and dynamics of the ecosystems at such sites.

For certain relatively well-known groups of organisms, or those of great actual or prospective scientific or economic importance, accelerated regional surveys might be carried out. Groups to be selected for such studies might include the flowering plants, vertebrates, and possibly certain groups of insects, such as macrolepidopters or beetles. In these groups, the intensive study of particular local areas (Lemée *et al.*,

1975; Lovejoy, 1975; Dodson and Gentry, 1978; Duellman, 1978; Lamotte and Borlière, 1978; Croat, 1979) is of special importance and ought to be encouraged along with other broad-scale inventories (Gentry, 1978). For many other groups of insects and other organisms, it does not seem possible to contemplate bringing a world survey to a reasonable state of completion in the foreseeable future. Thus we would probably be better off to select a few representative projects and concentrate our resources on them, so as to come to know well the assemblages of insects and other organisms at particular places or in particular areas (Adis, 1977, 1979; Wolda, 1978). Janzen's studies of the entire insect fauna in Parque Nacional Santa Rosa, Guanacaste Province, Costa Rica, are illustrative of this sort of approach, as are the many studies by scientists associated with STRI, which have made Barro Colorado Island in Panama one of the best-known portions of the tropics (Croat, 1979; Montgomery, 1978). Since the forest canopy is by far the richest zone for arthropods, the canopy-fogging and mass-collecting experiments initiated in 1979 by C. Gene Montgomery and T. L. Erwin (Smithsonian Institution, personal communication, 1979) at selected sites in the Amazon Basin are of special interest.

In addition to the gathering of basic systematic information, inventory work in the tropics might include surveys of other features of interest, such as potentially useful secondary compounds in plants. Such compounds are of both theoretical and practical importance. If large numbers of tropical plants are to be tested efficiently and rapidly for such compounds, however, the kinds of laboratories envisioned by Janzen (1977b) must be established and, by international subsidy, made widely available.

How may we improve the quantity, quality, of focus of biological inventories in the tropics, beyond the obvious point that more funding is required? First, it is clear that success depends on the activities of institutions and of qualified personnel in the tropical countries themselves. The relatively few systematic biologists in temperate countries can do little more than make a limited contribution to our knowledge of tropical biota during the few remaining years during which part of this biota will be present in fairly good condition. Temperate countries can contribute substantially to the advancement of knowledge in the area of tropical biology by transferring funds, books, and equipment to institutions in the tropics and making available support of other kinds, by expediting the access of scientists resident in tropical countries to advanced training programs in their major tropically oriented institutions, and by carrying out programs that make their scientific collections and libraries more available. This in turn would help in establishing more-adequate means by which resident systematic biologists could contribute significantly to the understanding of the natural resources in their own countries. Projects carried out by scientists in foreign countries should be cooperative and should be carried out with the aim of enhancing the level of biological research in the host countries. Thus, when identification manuals are planned, it should be remembered that they are to be used during a 20-year period when perhaps a third of all tropical organisms are likely to become extinct; we shall need manuals that can be produced quickly for local use, rather than more-comprehensive works that might require decades for completion.

Second, the several dozen institutions in temperate countries that maintain large collections and libraries concerned with tropical organisms should seek means to select regional and other priorities to enhance the effectiveness of their efforts. Systematic biologists need access to the countries of the world, but political considerations often make such access difficult. Institutions with regional specializations should encourage other institutions to participate in their programs. At the same time, by sustained attention to the problems and opportunities of specific tropical countries, scientists in foreign institutions could play an effective role in developing the institutions and cadres of trained biologists in the countries they visit regularly; future multiplier effects would benefit all. These effects might be accelerated by the widespread production of accounts of the plants or animals of particular tropical areas—practical handbooks available to foresters, educators, politicians, and other concerned groups. There are many who have an urgent need for basic information of this kind.

Specialists traveling widely in the tropics may be able to gain unique, firsthand impressions of particular groups of organisms and may eventually be in a position to produce important generalizations about them. Clearly, however, many of the hypotheses that are regarded as important today will be outmoded tomorrow, whereas the results of current specimen collecting will be considered important and useful for purposes that we cannot imagine. Wider collecting should therefore be encouraged, both by specialists and by those who might accompany them. In most circumstances, systematists can expedite the learning process if they will work in teams representing two or more scientific disciplines. Thus we would expect that a team composed of entomologists and botanists would accomplish more in a given period than entomologists alone or botanists alone.

Because of the severe limitations on human resources, and the urgency of the task, it is recommended that priority be given to biological inventory in regions selected on the basis of probable rapidity of disap-

Inventory of Tropical Organisms

pearance of the biota involved, their intrinsic interest, and initial degree of knowledge, all considered in the context of political feasibility. Priority should be given by funding agencies to institutional projects that reflect a careful review of the goals of the institution, the criteria mentioned above, and the evidence for coordination with other institutions and investigators on an international basis. The impact of the project on the future enhancement of the capability of the tropical countries involved should be given special consideration in evaluating these proposals.

For the reasons just reviewed, we propose that national institutions, such as the U.S. National Science Foundation, Deutsche Forschungsgemeinschaft, the Swedish Royal Academy of Sciences, and the Consejo Nacional de Ciencia y Tecnología (CONACYT) in Mexico, support the production of such inventories, perhaps with the understanding that they would be joined by such international agencies as FAO and the European Science Foundation. The lack of knowledge is drastic, and it is unlikely to be reversed greatly by existing programs. For example, there are no current catalogs for 22 of the 25 orders of insects that occur in South America, and not a single species list of the ants of any one area in the Amazon Basin. There is no country in South America for which a reasonably complete flora or checklist published in the twentieth century exists. Flint's (1971) studies of aquatic insects in the Amazon Basin revealed that 53 of 55 species of the Order Trichoptera of which the obtained samples were undescribed, and a pattern of this sort is by no means unusual when tropical faunas are investigated in detail.

EVOLUTIONARY SYSTEMATICS

In addition to basic inventory activities, systematists must take the lead in determining whether principles already derived from temperate-zone examples apply to tropical organisms. For such parameters as the size of effective breeding populations, the prevalence of changes in chromosome number and structure as evolutionary mechanisms, and the importance of hybridization in evolution, we know little about any tropical group of organisms. We do not know in detail for any such group the patterns of internal or external barriers that separate species or occur within their populations (Gómez-Pompa, 1967; Lowe-McConnell, 1969; Whitmore, 1976; Soulé and Wilcox, 1980).

Keeping in mind that roughly two thirds of all kinds of organisms do in fact occur in the tropics, and that only a very few of them have ever been studied by any modern method, are we really justified in drawing the kinds of far-reaching evolutionary conclusions that are commonplace in our texts and reviews?

In any event, greatly improved biosystematic understanding of tropical organisms cannot await the completion of their basic classification. Simply obtaining rough-and-ready classifications for a significant proportion of tropical organisms before they become extinct will be a race against time. Therefore, to extend evolutionary theory to the tropics in a significant way, it will be necessary to select specific groups of organisms and to study them in detail. No one investigator could hope effectively to apply the many diverse techniques required for a comprehensive understanding of any group of organisms at this level. It is virtually mandatory, therefore, to think in terms of research teams working simultaneously and from many diverse points of view on a few carefully selected groups of organisms.

As an example of the kinds of choices that might be made, consider the flowering plants of the Amazon Basin. At a minimum, it would seem wise to select examples of canopy trees, epiphytes, and understory plants, and to include in each category one or more groups that have many diverse species and one or more that have few. Other selection criteria might include distribution with relation to the major habitat types of the basin, the relationship of the particular pattern of distribution to postulated refugia (Simpson and Haffer, 1978), and the kind of breeding system. Perhaps such a selection grid would dictate a decision to make a detailed investigation of about 10 groups of flowering plants with a total of perhaps 300 species in the Amazon region, a small number in comparison with the more than 30,000 flowering-plant species thought to occur there. Such outstanding and widespread tropical plant groups as the figs (*Ficus*) or palms (*Arecaceae*) might be especially suitable.

Another fruitful approach might be to select genera characterized by exceptional diversity, such as the species flocks of fishes in certain closed lake basins, Hawaiian *Drosophila* (Carson and Kaneshiro, 1976), or the tarweeds (Asteraceae-Madiinae) of the Hawaiian Islands. Such tropical groups offer some of the best examples of rapid evolutionary radiation, and their pattern of diversity, as now understood, suggests that relatively few genetic changes have resulted through pleiotropy and other developmental interactions in very large phenotypic consequences. A detailed analysis of the genetic factors that are operative in such groups of organisms, coupled with a comprehensive examination of features of their anatomy and physiology, is likely to prove of great importance in theoretical evolutionary terms.

We emphasize that such a conscious selection must be made, and

Inventory of Tropical Organisms

detailed investigations carried out on the groups selected, if we are really to have any idea of the evolutionary patterns exhibited by tropical organisms. Overall surveys of organisms are useful, but neither these surveys nor the kinds of relatively superficial biosystematic investigations that may be carried out from time to time by individual investigators can ever add up to a detailed understanding of the mode of evolution of any rich tropical group. The time is simply too short.

Once decisions have been made to study certain taxonomic groups, assistance can be marshaled through existing organizations and institutions to help collect and study these groups. It is of fundamental importance to this effort, however, that the knowledge gained from field studies be quickly transferred into the hands of the persons resonsible for the management and protection of biotic organisms within tropical parks and reserves. For example, two centers for wildlife-management training in Tanzania and Cameroon now have graduates stationed in many African countries (V. C. Gilbert, National Park Service, Washington, D.C., personal communication, 1980). The U.S. National Park Service has an exchange program with the College of African Wildlife Management at Mweka, Tanzania, and would provide an ideal framework through which to facilitate such transfer of information.

HISTORICAL STUDIES

The species richness of tropical forest ecosystems cannot be understood fully without a knowledge of their history. Local variation in species richness and gaps in distribution may reflect in part climatic and other changes during the Pleistocene. Ultimately the only way in which the possible historical component in these patterns can be analyzed directly is by the examination of Tertiary and Quaternary fossils. Studies already carried out have begun to provide insights into the relationship among species richness, local endemism, and modern patterns of distribution, both of communities and of individual species (Flenley, 1979).

With respect to the Tertiary record, much information is already available in the restricted files of companies engaged in exploration for petroleum. It would be highly desirable if accords could be reached whereby the records would be made available for the interpretation of the migrations and past distribution of the biota.

Quaternary paleontological records are derived mainly from recent unconsolidated mineral sediments and peats, which must be undisturbed to be of maximum value. Irrigation, cultivation, surface mining, and a variety of onshore activities are rapidly eliminating suitable deposits that until recently were abundant in the tropics. We recommend that a study be carried out to determine areas of significance that are especially threatened and that this study outline the steps to be taken toward the initiation of a program to obtain borings at critical threatened sites, as well as to preserve some of the most critical sites.

PRESERVATION OF GENETIC RESOURCES

Systematic biologists should also play major roles in identifying genetic resources of critical importance that should be preserved and in helping to create the means to preserve them (see Kemp et al., 1976; Frankel, 1977; Soulé and Wilcox, 1980). A general knowledge of the diversity of organisms, in a geographical context, will provide the best guide for the selection of reserves. As Myers (1979a) has shown, it will no longer be adequate for conservation purposes in the tropics to concentrate on a few spectacular species, or even on a few outstanding habitats. A prudent approach would instead concentrate on areas of maximum diversity, especially on those with a high degree of endemism (Lovejoy, 1979a). In addition, the expertise of systematic biologists is indispensable in identifying groups of organisms of potential economic value and in determining strategies for preserving their genetic diversity, whether this be in *in situ* reserves, seed storage banks (Thompson, 1976), tissue-culture collections, botanical gardens, or zoos (see Ashton, 1976b).

Some of the pertinent techniques and approaches to the problem have been discussed in detail (Frankel and Bennett, 1970; Soulé and Wilcox, 1980). A review of the importance of sampling genetic diversity for tropical forestry has been provided by Kemp (1978). Veprintsev and Rott (1979) argue for the formation of tissue banks, stored in deep freeze, for rare and endangered animal species; similar methods have been suggested for plants. Investigation of the potentialities of seed banking and tissue-culture collections should be accelerated, but it is clear that the establishment of natural reserves affords the best approach to the preservation of genetic diversity and is the only truly practicable one for many organisms (Iltis, 1972, 1974; Miller, 1975; Jacobs, 1977; Matthew, 1978; Whitmore, 1980). In this connection, Myers (1979a) has provided estimates suggesting that a relatively high proportion of tropical diversity could be preserved if about 10% of the biome were set aside in well-chosen reserves. Whether these estimates are literally true or not, the effort appears well justified.

The practical significance of research on genetic diversity is especially important for maintaining the diversity of domesticated plants and animals. Many are being cultivated in rich diversity today, but their

Inventory of Tropical Organisms

diversity is continuously being eroded by the deliberate and continuous introduction of new, improved varieties and by the breakup of the aboriginal and peasant cultures that husband them. The subsidization of traditional agriculture *in situ* in indigenous, cradle regions, at least one for each major crop, which would amount to the "freezing" or preservation of the diverse aboriginal genetic landscape, should certainly receive consideration (Iltis, 1972, 1974). For example, the importance of manioc, both as a staple food and as a source of energy, is enormous; yet the number of cultivars that exist can only be estimated, and little information is available concerning many of them.

The virtually exclusive concentration by such international agencies as FAO on a few selected crops already known to be of value must be supplemented by much wider use of additional crops if world food stability is to be attained.

The significance of tapping aboriginal knowledge of wild forest resources is great. Local knowledge of plants will often be the first clue to recognition of those that will be useful in pharmacology. There is great potential for new forest products, oils, medicines, perfumes, insecticides, timber, and edible fruits that might be brought into domestication but are now known only by the aboriginal peoples who discovered them and who continue to use them as an integral part of their culture. The tree *Brosimum alicastrum* is an example of a wild plant of great potential utility. Studies concerning the past use of this tree by the Mayas and its great potential as a contemporary source of food have been carried out by the Instituto Nacional de Investigaciones sobre Recursos Bióticos in Mexico. If the widespread cultivation of this tree could be implemented on a larger scale, it would make a significant contribution to food production. It should also be noted that aboriginal peoples use and selectively encourage populations of certain animals, such as turtles (e.g., *Podocnemis*) and capybaras (*Hydrochoerus*) (Ojasti, 1968, 1971; Ojasti and Medina, 1972), as sources of protein. The role of controlled populations of these animals in human nutrition clearly deserves attention. Such practices are, however, being lost rapidly, and they should be studied while they are still remembered.

CRITICAL AREAS

We recommend that emphasis in collecting over the next 5 to 10 years be given to certain areas of the tropics that, because of their great biological diversity, high levels of endemism, and the rate with which their forests are being converted to other purposes, seem to demand special attention. Areas now largely untouched and likely to remain so for a relatively long time, though needing study, might serve as base lines against which to monitor further changes. The critical areas we have selected are as follows:

- *Coastal forests of Ecuador* The moist forest along the base of the Andes in Ecuador, with associated vegetation types, is extraordinarily rich in endemics even though only small areas still exist (Dodson and Gentry, 1978; Gentry, 1978). More than 50 new species of higher plants have been described during the past few years from the Río Palenque station (1.7 km^2), one of the few remaining areas of tropical moist forest in western Ecuador. The study of this forest should be pushed on as vigorously as possible. Priority attention should likewise be paid to the poorly known, but somewhat less threatened, area that extends northward through the Chocó of Colombia to the Darién in Panama.
- *The "Cocoa region" of Brazil* This zone occupies the southeastern extension of the state of Bahia, between the Atlantic coast and 41°30'W longitude and between 13°00' and 18°15'S latitude, as well as a small area near Linhares, farther south in the state of Espírito Santo (Mori and Silva, 1979). It covers about 100,000 km^2. Within this region only a small amount of the forest remains, and aside from a few reserves and parks, the rest will be gone within a decade (S. Mori, New York Botanical Garden, personal communication, 1979). It is estimated that the roughly 2 million people who inhabit it will double by the first part of the twenty-first century. The region is extremely rich in endemics, many poorly known, and novelties are being discovered with every new collecting effort.
- *Eastern and southern Brazilian Amazon* The forests of these areas are rich and diverse and are rapidly being destroyed (Aubréville, 1961). A number of the refugia that have been postulated on the basis of distribution patterns are located in the forests that are being decimated most rapidly. Especially in areas of relatively good soil, such as parts of Rondônia, forest farmers are permanently converting largely unknown lowland forest to other purposes (NRC, 1980). The area of the Belém forest refugium in the south of the state of Pará is an important center of endemism in which the forest cover is being removed rapidly.
- *Cameroon* Cameroon Mountain is the center of a moist forested area extending into adjacent Gabon and to the vicinity of the Cross River in southeastern Nigeria, including the Oban Hills. This area is characterized by high endemism and rich diversity (Letouzey, 1968; Brenan, 1978). The speed with which these forests are being destroyed (Myers, 1979b), and the fact that they are the most species-rich of western Africa, commends them for priority in study. The "Western

Block forests" of Sierra Leone, Liberia, and the Ivory Coast likewise deserve priority attention for similar reasons.

- *Mountains of Tanzania* Because of the richness of the plant, insect, frog, and mammal endemics of the moist tropical forest outliers of the Usambara, Nguru, and Uluguru hills of Tanzania and their associated ranges, and the related montane forests of Kenya, and the speed with which these forests are being destroyed, these areas urgently need study (Polhill, 1968; Brenan, 1978; NRC, 1980). In addition, the dry evergreen coastal forest that extends from southeastern Kenya to northern Swaziland, and exist as a mosaic in deciduous woodland, are rich in endemics and are being decimated throughout the area.
- *Madagascar* The highly endemic flora and fauna of Madagascar (Humbert, 1927; Richard-Vindard and Battistini, 1972; Leroy, 1978) are being destroyed rapidly (Rauh, 1979). Although the population in mid-1979 was estimated at only 8.5 million people, it is expected to double within 27 years; this contrasts with an estimated population of 2 million people in 1900. The rain forests of eastern Madagascar are of special interest biologically, were never very extensive, and are being destroyed rapidly. Madagascar represents a museum of the Cretaceous and Paleocene biota of Africa (Raven and Axelrod, 1974), and the detailed investigation of this biota while it still exists is a matter of the highest priority for systematic biology.
- *Sri Lanka* About a quarter of the angiosperm flora is endemic; most of these endemics are confined to the wet zone. The forests of this zone have been so extensively destroyed, and population pressures are so extreme, that the wet forests of Sri Lanka demand prompt attention. An exceptional concentration of endemics occurs in the coastal hills of the extreme southwest.
- *Borneo* The lowlands of Borneo, especially those of Indonesian Borneo, are the richest and most threatened block of tropical lowland forest in the Malesian area (Jacobs, 1979, and personal communication, 1979). For angiosperms, there are some 10,000 species, with 135 known endemic genera; a third of the species are trees. Borneo is the main center for Dipterocarpaceae (about 260 species), which are of great economic importance. Virtually all genera of fruit trees that are cultivated in Indonesia, including *Artocarpus, Citrus, Durio, Garcinia, Eugenia, Mangifera, Musa, Nephelium,* and *Salacca,* are well represented in the wild flora of Borneo. In addition, there are high concentrations of species in the Apocynaceae, Asclepiadaceae, Connaraceae, Dioscoreaceae, and Menispermaceae, which are extremely promising as sources of new compounds of high potential value as drugs. Yet even the major patterns of distribution and local endemism for flowering

plants are not well known for Borneo, except in a few districts. The lowland forest everywhere is being destroyed at such a rate that little will be left by the year 2000. Borneo apparently deserves more-urgent attention than any other part of tropical Asia.

• *Celebes* The lowland biota of Celebes is still rather poorly known, and the forests are apt to be destroyed rapidly, given the existence of transmigration and colonization schemes, widespread letting of timber concessions, and large-scale nickel mining. The forests where the greatest richness and diversity occur are in the central part of the island. The two southern peninsulas of Celebes possess relict forests of exceptional interest, some on ultramafic rocks. Celebes is the least well collected section of Malesia (Prance, 1977).

• *New Caledonia* Nickel mining and cattle raising are steadily eating into the vegetation of New Caledonia (Raynal, 1979). This island, which separated from Australia–Antarctica some 80 million years ago, is inhabited by many groups of plants and animals that occurred in Australia at that time (Raven and Axelrod, 1972); relatively few of these have been investigated in detail. New Caledonia contains the most archaic assemblage of flowering plants on earth, and about 90 percent of its flora, estimated at 3,000 species, is endemic.

• *Hawaii* There are no specialists who can deal with many of the largest groups of invertebrates in Hawaii. The last complete flora of the islands was published in 1888, and no comprehensive guide to the native and introduced plants is available. The destruction of the Hawaiian environment is proceeding with great rapidity, and consideration has for some time now been given to schemes to replace native forests with plantations of such trees as Mexican ash and Australian *Eucalyptus* so that the state could be spared the necessity of importing lumber and pulp from the mainland. Such exotic animals as pigs and goats are ravaging the vegetation, even within the national parks and supposedly conserved areas. The Hawaiian Islands are one of the great natural laboratories of evolution in the world, and government agencies (federal and state) should work to conserve the biota and provide adequate funds to study it.

4 Studies of Selected Tropical Ecosystems

We have concluded that the best way to develop a comprehensive understanding of how tropical forest ecosystems operate is to coordinate studies at a few sites that will receive detailed and continuing attention. It is necessary to focus on a few such sites because of the considerable expenditures required and the limited number of trained scientists who could reasonably become involved. Such concentration is intended to provide results in depth. We do not intend, however, that the excellent projects of this general nature now under way throughout the tropics be impaired in any way. Results from those ongoing projects will illuminate those derived from the major ecosystem studies recommended here. What we are suggesting is a major new initiative at several tropical sites; at each of these sites we hope to gain a deeper understanding of ecosystem structure and functioning than is available anywhere in the tropics. We have concluded that major expenditures for facilities, permanent staff, and long-term ecological monitoring coupled with ongoing research at a few selected sites in the tropics will contribute profoundly to the progress of tropical ecology as a science and accelerate fulfillment of human needs.

GOALS AND GUIDELINES

We want to know how natural systems operate in processing and controlling resource flows, and we want to be able to predict the effect of modification, by management schemes, of the temporal and spatial distribution of ecosystem resources. An understanding of the structure and functioning of an array of natural and modified tropical ecosystems will provide the basis for predicting the consequences of different modifications of the resources of any given tropical ecosystem type and for devising management plans to optimize sustainable resource productivity. Our aims, therefore, are congruent with those of Project 1 of UNESCO's Man and Biosphere Program (Anonymous, 1974b, 1976; Brünig, 1977; di Castri and Hadley, 1978).

We propose two guidelines for the study of tropical ecosystems that have not been incorporated generally into past studies of tropical ecosystems. First, each site must encompass a range of primary and modified ecosystems, including cut-over forests, naturally reforesting clearings, tree plantations, and various agroforestry systems. Second, research on the ecosystem must include an analysis of the roles of individual strategy types in processing resources and sustaining productivity. Ecosystem studies in the past have frequently focused on energy-flow considerations that have tended to reduce the biological diversity to large groupings in relation to their food-chain position. Similarly, watershed approaches have been concerned with inputs and outputs of water and nutrients and their mediation within the system. We propose here that both the energetic and watershed approaches be utilized in the study of tropical systems but that they be meshed with a detailed view of the biology and ecosystem roles of the resident population of organisms. A model for such a study is found in Bormann and Likens (1979).

If these goals are to be realized, long-term studies are necessary. For example, it is now recognized that nutrient- and water-cycling studies on an ecosystem scale must be carried out for decades to give definite answers to certain important categories of questions (Likens *et al.*, 1977). Meyer and Likens (1979), analyzing the long-term records of the Hubbard Brook ecosystem project, concluded that such records must be kept for varying but often substantial periods of years if they are to be used to determine whether the ecosystem is aggrading or degrading. Short-term variations, such as those of rainfall and tree fall, can give misleading results if the ecosystem is monitored for only a few years. The dramatic results at the Coweeta Hydrologic Laboratory regarding the effects on water yield of changes in species composition from hardwood to pine required decades of study (Swank and Douglass, 1974). Studies of redevelopment of ecosystems after clear cutting indicate that decades are required before hydrology and biogeochemistry return to precutting levels (Bormann and Likens, 1979). If significant results are to be achieved, periods of similar length will be required for the studies

Studies of Selected Tropical Ecosystems

recommended here. This conclusion underscores the need for long-term stability of funding and the creation of suitable management schemes for the continuing long-term research and monitoring at the sites selected.

The high-priority research topics identified in this chapter would, if imaginatively explored, provide a greatly improved understanding of the dynamics of tropical forests. Knowledge gained would be useful in solving a variety of practical problems, ranging from those encountered in designing and managing ecological reserves to those encountered in designing agroecosystems. More specifically, the goals would include:

- Obtaining information about the adaptive responses of biota to some of the richest terrestrial environments of the world before the opportunity for study is lost.
- Identifying ecosystems (and elements therein) that are found to be most urgently in need of conservation and preservation by reason of (1) the need to prolong the period available for studying the ecological basis of the environment and (2) the need to store valuable germplasm *in situ*.
- Providing an ecologically sound foundation for assessing and managing the secondary ecosystems that cover most of the tropics.
- Utilizing natural regenerative capacities to restore degraded soils and improve watershed management.
- Finding ways to use the potential productivity of tropical ecosystems to serve the needs of mankind without jeopardizing their productive base.

Traditional approaches to the study of complex ecosystems have answered different questions about them. Included are nutrient flows and cycling, species richness, ecosystem energetics, and ecosystem stability. The research topics that we have selected are those that we consider most relevant to the goals discussed earlier in this chapter. We have given high priority to topics whose applications to important management problems are most obvious. People are engaged in agriculture in the vicinity of all the sites identified for detailed ecosystem studies. Consideration of their agricultural practices in the context of the kinds of studies proposed, conducted in coordination with agricultural scientists, would be both instructive theoretically and useful in formulating productive agricultural systems for the respective regions under consideration.

A number of recent successes and promising beginnings have been reported in this field. We shall mention just a few as examples.

Multiple cropping in many tropical areas has been shown to produce higher yields (Bradfield, 1972; Innis, 1972; Trenbath, 1974). In Nigeria, Bede Okigbo of the Institute of Tropical Agriculture is studying, with considerable success, possible uses of native trees and shrubs for human and animal food; his work may have wide applicability elsewhere. In Venezuela, Zucchi (1975) has suggested that the pre-Hispanic system of raised fields is more productive than the European style of cultivating large, flat surfaces. Likewise, the research efforts of the Colegio Superior de Agricultura Tropical (CSAT), State of Tabasco, and the Instituto Nacional de Investigaciones sobre Recursos Bióticos (INIREB), State of Veracruz, Mexico, aiming at the introduction of chinampalike raised fields in southern Mexico (Gómez-Pompa, 1978; Gliessman, 1979; Orozco-Segovia and Gliessman, 1979) and of systems of irrigation in the highlands of Chiapas (see Matheny and Gurry, 1979) are well worth pursuing. Analogous methods of terracing and irrigation have been used traditionally in areas of Southeast Asia and probably could profitably be applied elsewhere. In general, the search for alternatives that employ and support more families than, for example, beef production is important and should be actively pursued. These and other methods should all be examined in terms of energy relationships and other ecosystem parameters, and the results should be disseminated as widely as possible for the benefit of all. In this respect, the First Conference on International Cooperation in Agroforestry, held in Nairobi, Kenya, in July 1979, clearly signaled an institutional revival of interest in the practice of shifting cultivation and other traditional practices world wide, in connection with their wider application and improvement. The systems used are extremely varied, and we have but limited understanding of their dynamics (see Bergman, 1974; Ruddle, 1974; Smole, 1976). An in-depth understanding of them is fundamental to the application of ecosystem research to the development of tropical agriculture in the future.

SELECTION OF RESEARCH SITES

In considering potential sites, we have given high priority to those that are representative of biomes that are in most immediate danger of extirpation, highly diverse, located in countries with a history of support for activities of this sort, and logistically convenient. This is also the objective of the international Man and Biosphere Project 8, which focuses on the development of a world-wide network of protected areas for long-term ecological research and monitoring (UNESCO, 1974). UNESCO and United Nations Environment Programme (UNEP) are currently developing plans for pilot environmental monitoring projects in selected biosphere reserves (V. C. Gilbert, personal communication, 1980). Thus the research sites recommended in this report should be considered for such pilot projects, if appropriate. We have in general used criteria similar to those advocated for experimental ecological reserves in a report prepared by The Institute of Ecology (TIE) for the U.S. National Science Foundation (TIE, 1977).

With these factors in mind, we propose sites in two New World wet tropical forest regions that occur on soils of widely different fertility, a site in Southeast Asia that would permit study of very rich forest in which representatives of the plant family Dipterocarpaceae are especially prominent, and a site in one tropical diciduous forest, a biome that is being altered even more rapidly than tropical rain forest. This is taken as a minimum number of sites at which to begin these investigations.

Several other kinds of sites are appropriate for detailed ecosystem studies of the kind we envision. Puerto Rico and Hawaii are important as two different kinds of island ecosystems, one more or less continental and the other oceanic. The great savannas of East Africa, with their impressive and diverse herds of grazing and browsing mammals, constitute an outstanding tropical biome. Other grassland areas, such as the llanos of western Venezuela and eastern Colombia, the cerrados of Brazil, and the savannas of West Africa (which have been well studied by generations of ecologists), also command attention. Continued and increased attention should be given to the Barro Colorado Island field station of the Smithsonian Institution, which has perhaps the most comprehensive inventories of any tropical area. A limited monitoring program of physical and biological parameters in a single watershed on this island was initiated in 1974. Ecosystem-oriented studies should continue in all these areas and will contribute significantly to our understanding. Nevertheless, we concluded that the four areas mentioned in the preceding paragraph should have top priority. They are chosen for compelling reasons.

Infertile New World Moist Lowland Forest

An excellent beginning has been made in understanding the dynamics of this ecosystem type. In 1974, the Institute of Ecology, University of Georgia, in conjunction with the Centro de Ecología, Instituto Venezolano de Investigaciones Científicas, Caracas, and the Max Planck Institute für Limnologie, Plön, Federal Republic of Germany, initiated an ecosystem project near San Carlos de Río Negro in southern Venezuela to test some of the hypotheses about tropical rain forest productivity (Klinge *et al.*, 1977; Medina *et al.*, 1977; Herrera *et al.*, 1978b; Jordan and Medina, 1978). The investigations at this site, conducted under the aegis of UNESCO's Man and Biosphere Program, have been confined largely to the dynamics of nutrient cycling. San Carlos is hard to reach except by air. The committee decided that a lowland forest site on infertile soils should be selected elsewhere, preferably on the *terra firme* of the central Brazilian Amazonia, and that the results obtained at San Carlos should be applied there. Certainly selected efforts should be continued at San Carlos, and the site should be maintained for these purposes.

Fertile New World Moist Lowland Forest

An excellent site is available at Finca La Selva, a site of some 730 ha in the Atlantic lowlands of Costa Rica. Owned by the Organization for Tropical Studies (OTS), a consortium of 26 universities and research institutions in Costa Rica and the United States, La Selva has been the site of detailed studies for more than 25 years (Frankie *et al.*, 1974a, 1974b). More than a hundred papers have been published on various aspects of the systematics and ecology of the plants and animals of La Selva. Initially this reserve was the locus of studies on plant ecology and succession by L. R. Holdridge, G. Budowski, and their associates, and later a primary site for OTS activities and hundreds of diverse studies in evolutionary biology. With the decision of President Rodrigo Carazo of Costa Rica (June 18, 1979) to extend Parque Nacional Braulio Carrillo to the borders of Finca La Selva, there is now available for long-term study an unrivaled elevational transect of diverse forest extending from 45 to more than 2,900 m above sea level. This extensive park of 50,000 ha encompasses most of the habitats characterisitc of the Atlantic slope of Central America. Much disturbed and cultivated land immediately adjacent to the La Selva site is available for experimental manipulation, and the area appears ideal for the sort of long-term monitoring and ecosystem study we envision. The development of the La Selva Field Station as a major center for long-term ecological research to accommodate 50–60 persons has the full endorsement of the OTS and of the government of Costa Rica, the latter expressed in a letter of September 17, 1979, from President Carazo to President Jay M. Savage of OTS.

La Selva and the extension of Parque Braulio Carrillo illustrate clearly the way in which knowledge about an area can lead to its preservation, with consequent opportunity for obtaining further knowledge in the future. The large number of studies conducted in this area, and in Costa Rica generally, have had an obvious and important impact on the enlightened conservation policy of the country, which in turn will permit greater flexibility in future choices for development.

Southeast Asian Lowland Forest

A meeting of biologists from throughout tropical Asia was held in Bangkok, Thailand, in September 1979, under the auspices of our committee, to consider the choice of a research site in the region. It was decided that Gunong Mulu National Park in Sarawak is clearly outstanding on scientific grounds. This area probably constitutes one of the most diverse sites in the lowland tropics, with an annual rainfall exceeding 4,000 mm and no monthly mean less than 100 mm. It extends in elevation from 100 to 2,500 m and includes large areas of primary forests on podsols, gleys and peats, yellow podsolic and latosolic soils, and limestone. Mulu was recently the object of a study by more than 70 scientists participating in a year-long expedition of the Royal Geographical Society of London. Results were presented in a symposium in London in September 1979.

The committee selected Mulu as the ecosystem site of preference in tropical Asia and recommended that negotiations be initiated to secure it as an international rain-forest site for long-term studies. Although access is by plane and river at present, a major east-west highway is scheduled for completion by 1985 and will extend through the northern plains of the park. If for some reason Mulu cannot be established as a major ecosystem site, sites in Sumatra or Thailand might be considered.

New World Deciduous Forest

The Estación de Biología at Chamela, Jalisco, Mexico, which belongs to the Instituto de Biología of the Universidad Nacional Autónoma de México, has been selected as an appropriate site for long-term ecosystem studies of tropical deciduous forest.

This well-protected site of about 1,600 ha is representative of a biome distributed from southern Sinaloa in Mexico to central Costa Rica, all along the Pacific Coast slopes, from sea level to altitudes of up to 500 m, and is similar to areas of cerrado and cerradão in Brazil. It is also represented in parts of Africa and Asia. This biome is undergoing extremely rapid deterioration throughout the world under the impact of rapidly growing human populations and, among other factors, the demand for beef, the latter particularly in the Western Hemisphere.

The site at Chamela has a forest composed of neotropical elements, most of them ultimately of South American origin, of relatively high diversity (some 90 woody plant species per hectare) growing on sandy, shallow granitic soils. Average annual precipitations are in the order of 1 to 1.2 m, concentrated in only 5 to 6 months of the year (namely June to November), the rest of the year being almost completely dry.

Several studies have been under way since 1973 at the Chemela site, both detailed studies of the main forest components (e.g., demographic studies of *Cordia elaeagnoides*) and whole-community studies, such as those of temporal and spatial patterns of leaf-litter production (now on its fifth year) and leaf-litter decomposition rates, analyses of mineral content of rainfall, throughfall, and leaves of the main plant species and of growth rates of some of the more important trees of the forest, studies on aging of some species, and recently a study on the impact of herbivores on plant productivity of the forest. Basic studies on the flora and fauna of the biome have been done in different degrees of detail and are intended to be completed by 1981. Several theses resulting from these studies are being prepared and will be published in late 1980.

The Estación de Biología Chamela is accessible all year by paved road. It has housing and laboratory facilities that will be enlarged during 1980 and 1981 to accommodate some 25 to 30 visiting scientists in addition to the permanent research staff of the station. Electricity (and a back-up generator) and radio communication are available.

It is to be hoped that analogous studies may be carried out at Sakhaerat, Thailand, an area of dry deciduous and dry evergreen forest with excellent facilities with about 1.2-m mean annual rainfall and a 6-month dry season. Extensive studies have already been conducted at Sakhaerat that in many ways will be directly comparable with those under way and envisioned for the future at Chamela.

OBJECTIVES OF THE TROPICAL ECOSYSTEM RESEARCH PROJECT

We envision a need at each station for long-term ecological research in order to provide basic background information against which other studies would be compared. We also assume that biological inventories would be especially intensive at and around each site and that studies related to freshwater habitats, soils, and human populations would be carried out in the vicinity whenever possible. It would be highly desirable if the sites selected for intensive investigation could be adjacent to or, better, surrounded by, MAB biosphere reserves and parks, so that research at them could be continued far into the future.

The physical environment is the prime independent variable to which biological interactions need to be related. Input from the physical environment is then modulated by living organisms and influences their adaptations and the ways in which they interact with one another. In general, the physical measurements we recommend at each site are those summarized in a series of reports on long-term ecological research prepared for the U.S. National Science Foundation (National Science Foundation, 1977, 1978; The Institute of Ecology, 1979). These measurements should include determination of macroclimatic data, geological mapping, and soils and sediments mapping, all of which should be carried out as part of the initial background work at each site. Subsequently, a continuous program of monitoring physical variables should be developed, providing estimates of central tendency, short-term variability, and extreme conditions for each parameter at several diverse stations within each site. The methods and recommended measurements of physical parameters and of chemical, terrestrial, atmospheric, and aquatic measurements, as well as suggested uniform procedures for storing samples for further investigation, are summarized in the reports cited above. We want to emphasize the extreme urgency of initiating such studies in the tropical regions of the world, where the environment is changing much more rapidly, and with even more far-reaching potential consequences, than in the United States, on which the reports mentioned are focused.

For each area we need maps of the surface geology, topographic features, and soils of the region. Because of the important roles that rare catastrophic events may play in ecosystem properites, it is important to make detailed measurements following any such events to determine changes in all physical and chemical parameters.

An important method for integrating data on diverse ecological parameters is the small-watershed technique (Likens *et al.*, 1977; Bormann and Likens, 1979). The utilization of small watersheds as basic units of ecosystem study provides a meaningful way of bringing together data on hydrology, biogeochemistry, weathering, and biological activities within the ecosystem. Where feasible, it would be desirable to employ this technique.

Because of the importance of being able to compare measurements made in different locations and at different times, standardization of measurement techniques is vital. Measurements should allow for separation of unidirectional and cyclic changes by time–series analysis and detection of any time lags in reponse to external influences. It is of great importance that the sampling plots be permanent. Studies of these kinds have generally been rare in the tropics, but there are a few examples (Windsor, 1976).

RECOMMENDED SUBPROJECTS WITHIN THE ECOSYSTEM RESEARCH PROJECT

An integrated view of ecosystem dynamics can arise only from investigation of a number of interrelated focal areas, here treated as subprojects within the overall scheme. The following subprojects are recommended:

- *Water and nutrient cycling* This subproject would be concerned with quantifying the temporal and spatial distribution of the physical resources within the natural and modified systems. It would include, in particular, water and nutrients in their various states in the atmosphere, soil, and biota.
- *Ecosystem energetics* This subproject would be concerned with quantifying the temporal and spatial distribution of energy within the natural and modified systems. It would entail measurements of components of solar radiation as well as of the productivity and standing crops of plants and of various plant parts and of the organisms that feed on them. It would assess the quality (both energy and chemical content) of plant components for various herbivore groups.
- *Physiological plant ecology* This subproject would be concerned with determining the mechanisms by which plants acquire and allocate carbon, nutrients, and water resources in tropical ecosystems.
- *Herbivory* The utilization by animals of living plant parts constitutes the largest junction in the transfer of energy from plants to other trophic levels in the ecosystem. Pollination systems constitute one important subset.
- *Higher-order food webs* The higher-level interactions, involving the influence of carnivores and parasites, play a major role in structuring ecosystems.
- *Dynamics of microhabitats and patches* Patchiness in space and time can trigger complex patterns of invasion, growth, and extinction within and among patches.
- *Soils research* Characterization of the soils in areas of undisturbed vegetation, and of the ways in which they are changed after the removal or modification of the vegetation, is essential to an understanding of ecosystem functioning.

It is anticipated that all these studies would be carried out at each of the ecosystem sites, ideally in a watershed ecosystem, although the emphasis would vary in relation to the questions being asked and the opportunities locally available.

Water and Nutrient Cycling

A major comparison among tropical ecosystem studies is between the relatively nutrient-rich soils characteristic of recent volcanic and some riparian situations and the nutrient-poor soils characteristic of regions having ancient, deeply weathered bedrock or oligotrophic sediments. There are, of course, soils of intermediate fertility, but there are large regions with either ancient or fairly recent soils. Although striking differences are to be seen in the ecological communities found under those contrasting conditions, some results from undisturbed ecosystems suggest that the rate of plant biomass production in the poorest tropical soils is about the same as it is in the richest (Ashton and Brünig, 1975). It is of great theoretical and practical importance to understand these relationships better.

To compare ecosystems or generalize about their nutrient cycles, the most important questions need to be posed. One important issue concerns species diversity and its relationship to disturbances; observed changes in species diversity may be important indicators of other changes. Another important question concerns the significance of the tightness of nutrient cycling. This is usefully measured by the recycling index (Finn, 1976, 1978), and it is an example of one way in which to begin to explore these problems. It is an index of the ratio of the amount of nutrients recycled in an ecosystem to the amount moving straight through. A recycling index close to zero means there is little recycling of nutrients in an ecosystem, and an index of one indicates total recycling. Although the use of this index tends greatly to oversimplify the situation studied, it does offer certain advantages.

In the oligotrophic forest of the Amazon Basin, recycling is very high in comparison with straight-through flow. Several mechanisms contribute to the high recycling, including (a) the mat of humus and roots on top of the mineral soil that rapidly absorbs nutrients released by decomposition or brought in by throughfall (Jordan and Stark, 1978; Stark and Jordan, 1978) and is a buffered system that maintains the soil water at a low pH, thereby inhibiting loss by microbial activity, especially nitrification (Jordan *et al.*, 1979); (b) epiphytes growing on leaves and bark that act as exchange columns for the nutrients in precipitation, and fix nitrogen, thereby increasing the nutrients available to the forest (Jordan *et al.*, in press); (c) the thick bark of some of the trees that inhibits leaching by stem flow (Jordan and Uhl, 1978; Jordan, 1979). The generally sclerophyllous leaves reduce nutrient losses because of their longevity, strong chemical and mechanical defenses against herbivores, and resistance to leaching of nutrients (Janzen, 1974b; Medina *et al.*, 1978). The surprisingly high productive capacity of the undisturbed Amazon ecosystem, when the infertility of its soils is considered, undoubtedly is due to the high internal nutrient recycling.

The recycling index also is applicable to the study of water budgets. A high recycling index for water has the same significance as a high recycling index for nutrients. Villa Nova *et al.* (1976) and Salati *et al.* (1978) have reported that in the Amazon Basin about half of the water that falls on the basin is transpired and again falls over the basin. The rest runs off to the ocean. Thus the recycling index is about 0.5. In contrast, Budyko (1974) has estimated that in much of the rest of the world only about 12% of the rainfall comes from ecosystems within the region.

Priority research needs in tropical agriculture deal with some aspects of these same problems. Indigenous agronomists and agriculturalists in the tropics have learned to some extent to manage ecosystems for subsistence by returning a part of the aboveground biomass, such as forest, back to the ground as ash to enhance its productivity for annual crop production, and this is followed by regrowth of the forest fallow (Pelzer, 1978). When populations attained levels above the capacity of the system to support, them, however, environmental deterioration was often correlated with nutrient loss (Deevey *et al.*, 1979).

Ecosystem Energetics

The amount of energy captured by green plants is dependent on the radiation impinging upon them as well as the amount of nutrient and water resources available. Annual solar radiation incident at the surface of the earth's atmosphere increases from the poles to the equator, but radiation incident at the ground or forest canopy in the tropics is modified and reduced by the greater thickness of the atmosphere itself toward the equator and by the prevalent cloudiness of the humid tropics. Actual incident radiation is measured at few sites in the humid tropics, yet this must be considered the ultimate determinant of productivity. An understanding of the way in which this incoming energy is transferred and stored in nature, and its adaptive significance in terms of growth rates, reproductive potentials, defenses against predation and disease, temperature, and water balance, is also vital to the development of sustainable productive agriculture.

The relatively few available estimates of the primary productivity of tropical forests vary greatly. They are mostly based on inadequate and unreplicated samples or on extensive surveys yielding approximations the confidence limits of which are unknown; the detailed studies of small samples in lowland Malayan rainforests, extended by Kira and his co-workers to large stands through allometry, are a noteworthy exception (Kato et al., 1978; Kira, 1978; Yoneda et al., 1977). Many more determinations are needed to resolve the patterns adequately. There is now some doubt about whether natural tropical forests are indeed as productive as once believed, and more data bearing on this central question are needed.

Studies on productivity of tropical systems need a broader base, however. We need to know how the energy-capturing capacity of naturally disturbed and man-modified systems can be not only assessed but manipulated in the interest of preservation of natural ecosystems or others that can produce higher sustained yields of useful products.

We need to know total biomass accumulation rates for our experimental array as well as biomass decay rates to resolve the critical question of whether tropical forests or the modified variants are sources or sinks for atmospheric CO_2. A comparison of natural forests versus successional states would be invaluable in this context.

Although there are a number of important questions that can be answered by studying the productive components of tropical ecosystems in the traditional manner—that is, by considering the productivities of the basic units of producers, consumers, and decomposers—there is great merit in taking a more refined approach. It is only through understanding the productive capacity of individual species that decisions to favor one species over another can be made. This single-species approach is described further in the statement on the physiological ecology subproject.

Generally, studies of productivity lump all organisms into one or the other of the major trophic classes, and they often fail to consider seasonal variation in total production and also variation in the different parts of the same organism. It is such fine detail, however, that ultimately determines the entire structural and seasonal dynamics of an ecosystem. In studies of productivity of tropical systems, we need to know what types of new leaves, flowers, and fruits are produced, how many are produced, and when. These details control the numbers, kinds, and life cycles of herbivores and pollinators within a system.

Studies of primary production need to be extended beyond the usual measures to include estimates of allocations to reproductive efforts (nectar, pollen, fruits, seeds), extent of losses that are due to high nocturnal respiration by plants, and consumption of tissues by animals. The relative scarcity of frugivores in tropical forests on nutrient-poor soils may indicate that, although allocations of energy to wood and leaf production are sufficient to maintain growth rates similar to those existing on nutrient-rich sites, allocations to reproductive effort may be much lower. This may account for much of the difference between overall productivity and the part of productivity usually measured in energetics studies. By carefully integrating nutrient-cycling and energetics studies with those concerning utilization of plant tissues by animals, we can substantially improve our understanding of tropical forest dynamics.

One other dimension of the studies of the primary productivity of tropical ecosystems that needs considerable emphasis is the quality of the biomass produced. It has become obvious in recent years that plants produce an array of secondary chemicals that serve as protectors against various herbivores and pathogens. Thus, to predict how much energy is available for a particular consumer, we must know more than the amount of biomass produced, its energy content, and the seasonality of production; we must also know the chemical nature of the product. A study of the secondary chemistry of plant products is an essential part of the energetics subproject and of the regeneration part of the physiological ecology subproject.

Physiological Plant Ecology

Knowledge of the physiological ecology of tropical plants is necessary for our understanding of the functioning and management of natural tropical ecosystems as well as plantation and agroforestry systems. Physiological ecology is concerned with determining how organisms acquire and allocate resources in specific environmental complexes. Such information tells us the principal adaptive modes of any organism, and it can also be used to predict species interactions (e.g., seasonality of allocation in a producer indicates resource-availability patterns for herbivores).

Although some information is available concerning the ecophysiology of tropical crops (see Alvim and Kozlowski, 1977), our knowledge of the acquisition and allocation patterns utilized by tropical plants is so limited that virtually any research effort will result in valuable new information (Sarukhán, 1979; Mooney et al., 1980). However, some simple guidelines can direct us to the highest-priority research directions. First, organisms utilizing a given resource base and having comparable life forms—e.g., canopy emergents or understory herbs—may have similar generalized resource capture and allocation patterns. Hence initial research programs might have a life-form group focus rather than a taxonomic focus. Concentrated studies on representative life-form types can give us a broad understanding of the principal adaptive possibilities. An example is the significance of crown–stem diameter ratios to growth rates in shade-tolerant and shade-intolerant tree species. Second, research obviously should concentrate on acquisition and allocation processes related to the resources most limited in the study environments. For some tropical humid systems, this means a study of the mechanisms in the capture of nutrients and light, whereas in others, such as in savannas and tropical deciduous forests, mechanisms related to efficient use of water must be studied, as well as nutrients. Finally, a focal point for priority research should be the life-history event most critical to establishment and regeneration.

The specific research priorities recommended for the study of the physiological ecology of tropical plants are as follows: nutrient balance, regeneration potential, and carbon dioxide exchange, as well as the establishment of improved research facilities.

Nutrient Balance One of the highest-priority research needs is the study of the nutrient balance of individual species representative of the total array of plant-life forms found in sites of differing nutrient availabilities. The balances in the plant–soil–microbial association needs study, with an emphasis on N, P, S, Mg, and Ca.

The low soil-nutrient stores and high acidity of most moist tropical regions have presumably led to unique adaptive features of the plants that inhabit these regions. Are there plants with

- unusual fixation or nutrient-uptake capacities?
- mechanisms that lead to tight internal recycling?
- high tolerances of elements such as Al or Mn?
- novel nitrogen-fixation associations?

Answers to such questions could help in developing management programs for tropical moist regions and provide information that would help in making predictions about the consequences of specific disturbance events. In addition, such information would provide guides for selection for potential agroforestry species for soils with low P, high Al^{+3}, and so on.

In tropical areas with more-limited moisture (e.g., deciduous forest and savannas), the following nutrient relationships need study:

- Nutrient allocation to storage in relation to fire.
- Interactions of nutrient and water balances.

Regeneration Potential An important topic that needs considerably more attention concerns the factors involved in the regeneration potential of tropical plants. Although considerable work has been done on the populational aspects of the reproductive biology of tropical plant species, information on the physiology and adaptive morphology of reproduction is seriously inadequate. This information is of basic interest as well as of central importance in reforestation and in the analysis of the impact of perturbations. We need information on:

- The environmental triggers for reproduction; the relationship of resource levels to the timing of flowering and dormancy.
- Regenerative capacity of different growth forms; the types and functions of reserve meristems.
- Adaptive significance of various morphological seed types; nature and extent of seed reserves; consequences of the lack of dormancy in seeds of many species of the tropical evergreen forest.
- Carbon, water, and nutrient balance of seedlings.
- Mechanisms of protection of seeds and seedlings against predators.
- Methods of establishing mycorrhizal associations.

Carbon Dioxide Exchange There is an urgent need for more-detailed information on the carbon dioxide exchange rates between the atmosphere and individual tropical plant types as well as for whole ecosystems and their components.

We have identified these specific research needs:

- Mechanisms for frugal carbon dioxide exchange in understory plants.
- Mechanisms for gas exchange of flooded evergreen species under anaerobic conditions (e.g., *várzea* in flood season).
- Mechanisms of acclimatization to changing light regimes (emergents).

In addition, there are more-general research needs having to do with the relationships among growth forms, photosynthesis and nutrients, water, and light resources. Concurrent studies of respiration and carbon-allocation patterns would provide an analysis of the total possibilities of water-, light-, and nutrient-use efficiencies. This information is central to an understanding of the controls on plant regeneration and distribution. Further, such information provides the basis for predictions of responses to perturbations and for developing management potentials and policies for various agroforestry systems.

Temperature Relationships A number of questions related to the interaction of temperature and tropical plants need study:

- The basis for and significance of differing chilling sensitivities of plants. This is particularly important in tropical highlands.
- The relation between respiration rates in highland species (which occur where the nights are cool) and species in lowland areas.
- The basis of the reported mid-elevational peak in ecosystem production in tropical regions (Fournier, 1969; Janzen, 1973b, 1973c).
- How the apparent lack of spatial and temporal thermal gradients affect the metabolism of tropical species, especially in relation to reproductive biology.

Research Approach A research approach combining an analysis of the responses of plants under natural conditions with an analytical phase to determine the biochemical, physiological, and structural basis for the observed responses (to be carried out both in the field and in the laboratory) should yield the most fundamental understanding of the adaptive characteristics of tropical plants.

A facility needs to be established that would satisfy the requirements for the analytical phase of this project. In particular, facilities must be available for the growth of experimental material, in addition to suitably equipped laboratories. Puerto Rico is the most promising site for such a facility from a number of points of view since laboratories and tropical research groups already exist there and a wide range of community types is available for study; these types extend from undisturbed nutrient-poor montane rain forest to a variety of plantations, and from monoculture (sugar cane) to mixed agriculture (coffee and small garden plots). Further, a strong ecological base has been provided by the pioneering Luquillo Experimental Forest study (Odum, 1970). This forest has been designated a U.S. Biosphere Reserve and an Experimental Ecosystem Reserve (The Institute of Ecology, 1977), making its use for these purposes especially appropriate.

It is anticipated that many of the detailed analytical studies requiring sophisticated laboratories for physiological ecology would take place at the center in Puerto Rico, with proper application to and input from the other designated sites (i.e., Chamela, La Selva, and others) involved in the tropical ecosystem study. The construction of hypotheses based on initial observations would make possible precise experimental studies at the other sites.

Herbivory

Living plants present a richness of food resources for animals and fungi. For convenience, plant parts can be grouped into wood and roots, leaf parts, floral rewards, and fruits and seeds. These resources and their subdivisions are sufficiently distinct in nutrient and defense content that they are often eaten by quite different groups of animals. Consumption of these different tissues has, in turn, different effects on plant fitness, harvestable productivity, competitive abilities, population dynamics, spatial locations, and other phenomena. Insects are especially important in this regard, but we hope that the study sites will contain enough of the larger vertebrates to enable us to examine the effects of their predation as well. The questions posed below, in addition to identifying intrinsically interesting ecological problems, should increase our understanding of patterns and processes of interest to integrated biological pest-control programs, plant-breeding programs, and the cultivation of phytochemically important crops. This knowledge will also be especially useful in attempts to predict the effects of the removal (e.g., of elephants, riceborers) and the introduction (e.g., of honey bees, black rats) of selected species and of manipulation of groups of species (e.g., removal of the phytophagous insects by heavy insecticide application).

To investigate these interactions, it is first necessary to determine the distributions of chemical and physical defenses of plants according to taxonomic units, seasons, life forms, health and age of plant, successional stages, habitat, type of tissue, and so on. Knowledge of these patterns provides the background against which important questions about patterns of utilization of plant tissues by animals can be framed and studied. Examples of some of the more important questions follow:

- What are the host specificities of herbivores with respect to such parameters as herbivore life-history stage, host health, population density, and habitat?
- What is the intensity of herbivory with respect to this heterogeneity?

Studies of Selected Tropical Ecosystems 79

- What are the responses of the individual plants and their aggregations (populations, "communities," etc.) to losses of tissue to herbivores?
- How do the natural history patterns of herbivores relate to resource heterogeneity?

The nature of answers to these and similar questions will also throw light on another important area: competitive interactions among herbivores and pathogens and their importance for population dynamics and species richness of both the herbivores and the plants upon which they feed. Examples of questions to be raised about these interactions follow:

- How do congeneric herbivores and fungi divide up seemingly homogeneous resources?
- How important is competition among herbivores belonging to widely separated taxonomic groups, such as insects, birds, and mammals?
- What determines patterns of seed fall around fruiting plants, and how are these patterns related to coevolution?
- What is (was) the impact on plants and on competing herbivores of the removal of single pollinators or complexes of pollinators or of dispersal agents, regardless of whether they were removed through human harvesting, natural loss, or past catastrophic events, such as Pleistocene mass extinctions?

Such questions need to be answered if the adaptive significance of ecosystem energetics in the tropics is to be understood. These answers will be essential if models are ultimately to be built that can predict the optimal balance of resource allocation between crop yield and chemical defense, and between monoculture and polyculture of increasing spatial or temporal complexity.

The pragmatic and theoretical exploration of the above questions will generate a richness of hypotheses, conclusions, and data that should be of great value to the other aspects of tropical biology, especially the field of ecological-evolutionary biology, and to reduce losses to herbivores resulting from our uses of tropical forests. Intensive study of these questions over a 5-year period, extended as necessary, will greatly advance this study of animal-plant interactions in and out of the tropics.

Some long-term monitoring of the following activities that is needed by the overall study will be useful to the herbivore portion as well:

- General daily rainfall, temperature, and other weather parameters designed at the least to reflect conditions in open clearing and tree falls, forest understory, and forest canopy.
- Leaf, flower, and fruit phenology for species of trees that are the focus of breeding-system and physiological studies. The phenological traits to be recorded should be worked out in consultation with other researchers using the same or similar data.

Whenever possible, the animal, fungal, and plant species chosen for intensive study should be those also being used for physiological, behavioral, morphological, and anthropophilic investigations. Arrangements must be made for prompt identification of all voucher specimens.

Higher-Order Food Webs

The fraction of ecosystem energy and materials flowing through carnivores is generally very small. Despite this, the influence of these animals on patterns of competition and species richness in ecosystems may be very great owing to the tipping of delicate competitive balances by even weak selective pressures. Therefore, studies of plants and herbivores should be integrated with ongoing work focusing on the higher trophic levels that are built upon these energetically rich interactions.

Interactions among carnivores and between carnivores and their prey are poorly understood in tropical forest ecosystems. So little is known of their natural history that the most interesting questions are difficult to pose. A major thrust, then, of this study should be to gather information on carnivores for the purpose of generalizing hypotheses about underlying processes. An understanding of carnivores is relevant to tropical biology for three conspicuous reasons. First, various carnivores may be useful in our efforts to depress densities of herbivores that compete with human use of plants. Second, the indiscriminate removal of carnivores may alter and usually increase herbivore densities that were important in population regulation, and thus ecosystem stability, in their unaltered state. Third, carnivores are often vectors of human and livestock diseases. We identify the following immediate areas of highly productive research:

- Why do the number of species of at least some groups of entomophagous parasitoids (parasitic wasps, tachinid flies) and the number of individuals per host species and per host individual decrease with decreasing latitude?
- What is the role of infection by disease organisms in shaping com-

munity structure, which is classically considered to be set by limiting-similarity competition processes (e.g., competitive displacement by congeners, character displacement, size ratios)?
- Does the removal of the conspicuous higher-order carnivores (e.g., big snakes, social wasps, eagles, army ants, spotted cats) result in recognizable changes in lower levels of the trophic web?

Dynamics of Microhabitats and Patches

Decades of work on simple ecological systems in the laboratory have demonstrated the importance of competitive exclusion and extinction among species exploiting similar resources. The fact that species coexist despite great overlaps in resource utilization suggests that the temporal and spatial patchiness of nature may be important in preventing competitive interactions from reaching the same result. In tropical forests, patchiness is due to factors such as microtopography, landslides, death of trees, activity of animals, and, on an ever-increasing scale, logging and clearing for agriculture. The responses of communities of plants and animals are expected to be different, the differences depending on the nature and scale of patches created by these disturbances (Whitmore, 1975). A thorough study of these relationships can provide both an understanding of the dynamics of "natural" forests and valuable imput for devising management plans for exploiting those forests with minimal disturbance to important biological interactions (Ashton *et al.*, 1978; Hubbell, 1979; Lovejoy and Rankin, 1979).

Research into patch dynamics might profitably focus on two important aspects of plant biology, plant-breeding systems and the adaptations of seeds and seedlings to the heterogeneous conditions created by various gap-forming processes. Efforts should focus on the relationships between these processes and the unusually favorable climatic conditions for germination and early growth combined with the large numbers of species competing for access to these opportunities.

Proposed Research Topics This committee proposes four promising lines of research:

- *Spatial heterogeneity of plant resources* It is impossible to manage an uneven-aged forest containing a thousand tree species profitably on a sustainable basis since the product cannot be standardized and concentrated. We do not know the effect that simplification will have on population stability, genetic variability, and, in particular, the balance between specialized herbivores and pathogens and their tree hosts. Simplification implies that some species that are now uneconomic will become rarer or become extinct; it would be dangerous to assume that none of these declining species is of value.

• *Temporal heterogeneity of plant resources* The irregularity of flowering and fruiting makes the availability of these resources unpredictable (e.g., ylang-ylang oil from *Cananga* flowers; cocoa butter from illipe fruits, *Shorea* spp.), a fact that has prevented many species from becoming major commercial crops. The phenology of leaf change must be understood because of its influence on litter production, nutrient cycling, and pest infestation.

• *Life histories of tropical trees* The temporal significance of regeneration, gap filling, mortality because of competition, rate of growth in height and diameter, "senility," and ultimate death of mature trees of tropical forests is known only in general terms (see Whitmore, 1975, for Southeast Asia). Despite this, life histories are of primary importance to an understanding of the dynamics of ecosystems and their susceptibility to management for the production of wood or other products. There is a need to acquire more of this information for a number of tree species representing different roles in the system.

• *Genetic variability of plant resources* It is an astonishing fact that the entire rubber industry in Southeast Asia was built on a single seed consignment of *Hevea brasiliensis*, Pará rubber, from a restricted locality between Santarém and Manaus, Brazil. Prolonged selective breeding in this species resulted in a 10-fold increase in latex production, but improvement reached a plateau by 1948. Subsequently it was necessary to return to the Amazon to establish a broader genetic base for this crop, and new commercial clones have now been developed. The research on *Hevea* has provided the technology to investigate the population genetics and breeding systems of other potential rain-forest tree crops in anticipation of future needs. It has also demonstrated the necessity for adequate genetic conservation.

Program of Study The committee proposes two programs of study. One program relates to undisturbed systems and consists of four items. Two of these items are studies of seedling establishment and analysis of genetic variability through controlled crosses. The others are as follows:

• *Detailed long-term quantitative monitoring of tagged individuals* Basic data needed for all species in environmentally uniform forest samples, which nevertheless include all phases of the dynamic cycle, are phenological observations of leaf change, flowering, and fruiting; determinants of sexual and asexual phases; bud and seed dormancy;

germination; flowering biology; identification of pollen and fruit vectors and an analysis of their foraging behavior and effectiveness; tests for self-compatability and incompatibility; and causes of flower mortality and dispersal.

- *Monitoring of selected species representing all phases of each major forest type* Observations on the degree of synchrony of phenological events; monitoring of population structure, growth, and mortality and their causes; analysis of genetic variability.

The other proposed program of study relates to modified systems. Objectives are as follows:

- Identification of those types of organisms that, because of their key roles in system dynamics or because of physiological or adaptive limitations in their regenerative capacities, present special cases of jeopardy in the face of the types of ecosystem modifications that are now commonplace in the tropics.
- Determination of the causes, sequence, mechanisms, and abiotic significance of the regeneration of tropical ecosystems and their key component organisms following human interventions, such as hunting and gathering, partial timber extraction, deforestation, cultivation, grazing, and burning.
- Identification of stimuli useful in initiating or accelerating the regeneration of tropical ecosystems.
- Development of techniques, through these projects, to tap the growth and regenerative capacities of tropical ecosystems for the sustained production of useful crops.

Soils Research

Soils are integral to many of the previous considerations in this study, ranging from their role in nutrient cycling to their interrelationships with human ecology and the use of tropical areas (NRC, 1972). Most tropical soils research is dominated by the overwhelming need to develop permanent systems of agriculture and forestry that will not result in soil deterioration. The wide variety of tropical soil conditions and the various fertility and erosion-control management problems associated with tropical development are sufficient to keep all available soils researchers busy for a long period. The priorities for research that we recommend are important because they were developed by understanding the biological interactions of still-intact soil–vegetation systems.

Tropical forests on the many soil types have maintained soil fertility through their many mechanisms of cycling elements from soil through vegetation and back again. Indigenous and subsistence cultures of people at low densities have frequently interacted with these soil–vegetation systems and have established long-lasting cultural practices that maintain nutrient cycles. Increasingly complex societies have tended to operate in ways less dependent on the original soil fertilities (Zinke, 1977).

The specific research priorities recommended for the study of tropical soils are as follows:

Site-Specific Studies Relating to Nutrient Cycling and Soil Fertility These studies are best carried out in conjunction with intensive ecosystem study sites, as specified previously. They will involve

- Local soil description, classification, and mapping on the study site and determination of the relation of the site to the regional soil mosaic.
- Characterization of the nutrient-storage properties of the soils at each study site in terms of the influence of the various plant species on soil properties, including the local variation at each site.
- Microbiological characterization of the soil environment of each site and the role of the microflora and fauna in soil-fertility maintenance.

Tropical Forest Soil Studies External to Sites Collation of existing knowledge of soils is needed in the context of ecosystem studies. Most soils knowledge has been gathered in an agronomic context, but it could readily be translated to the needs of ecological understanding by relating it to natural preexisting systems of vegetation and soils relations. Much information based on soils studies has already been gathered, but it is not readily available. This project involves the collection, aggregation, and synthesis of the existing knowledge of soils (disturbed and undisturbed) in tropical forest environments.

Relations of People and Soils Through Case Studies In the context of the world system, it is necessary to consider human populations, their densities in various environments, and modes of economic development, and the impact of these factors on tropical soils, especially in relation to conversion of forests to pastures, to fast-growing tree crops, and to coffee and alternative crops. In addition, studies of the effects of traditional systems on soil fertility will be necessary.

5 Tropical Aquatic Systems

When priorities for the study of tropical inland waters were assigned, likelihood of alteration, ecosystem type, and conceptual significance were considered jointly. The freshwater ecosystem work recommended here is geographically dispersed, and the concentration of effort at a few permanent sites is not recommended, as it was for terrestrial ecosystem studies. Studies of significant duration (5–15 years) are required, but the locations of such studies must shift to cover the geographic range. Sites that have been studied intensively for a decade would not simply be left untouched thereafter but would be incorporated into the long-range monitoring plans of the relevant country. It will also be possible to conduct long-term studies on the small aquatic systems at the permanent terrestrial field sites, but these systems cannot be given highest priority until the large, uniquely valuable, and rapidly changing systems are studied much more completely than has been possible in the past. Marine systems, including mangrove swamps, have not been considered to fall within the scope of our committee's task.

We concentrate here on studies of the greatest urgency. A more detailed list of sites potentially valuable for aquatic studies is given by Luther and Rzóska (1971), along with some information relevant to each site.

RIVERS

Tropical regions contain several very large river systems and many tributary systems or minor river systems. Significant proportions of these are at present essentially unmodified by human activities. Studies now can therefore reveal their character before watershed development of various kinds occurs. In the reasonably near future, major changes will occur in all the very large watersheds and in many of the small ones. Principal changes include watershed deforestation, impoundment for energy production and flood control, and industrial and urban pollution. These changes will occur at a more rapid pace in some parts of the world than in others. Priorities for study must be based partly on the degree and likelihood of change.

Very Large River Systems

The Amazon, Orinoco, and Zaire (= Congo) rivers drain large amounts of moist forest and at the present time are in nearly natural condition. All these rivers are very poorly known in an ecological or biological sense (Rzóska, 1978). Deforestation, hydroelectric development, and pollution from pulpwood processing, mining, heavy industry, and refineries are highly likely for the Amazon in the near future. The same is true for the Orinoco; here it seems that, in the immediate future, changes in water quality are most likely to result from industrial development at Ciudad Guayana, from extensive impoundment, and from alterations caused by clearing on the slopes of the Andes. The Zaire is probably not so imminently threatened by pollution and impoundment, but it will undergo considerable change in the near future if the current worldwide trends in deforestation persist.

The Nile and Mekong rivers are also very large rivers, but they have already been greatly altered. The Nile is impounded at several points (Rzóska, 1976), and the Mekong has a long history of watershed development. The Niger River might also be considered in this class of very large rivers but does not drain extensive tracts of wet forest under imminent threat of destruction.

Given this general perspective on the major river systems of the tropics, it seems wise to place highest priority on the study of the Amazon and Orinoco rivers and their main branches. We have assigned the Zaire a priority ranking just behind the Amazon and Orinoco.

Other major rivers are eliminated from designation as highest priority for research on grounds that they have already been seriously altered

Tropical Aquatic Systems

or do not face immediate alteration. This in no way indicates that these rivers are unimportant, but it does mean that the main emphasis in the next few years should be on recovering data that will otherwise be irretrievably lost.

Since the very large river systems incorporate vast numbers of small streams and minor rivers, guidelines are needed to provide the basis for decisions about the allocation of effort in the immediate future. Very small watersheds can probably be studied in the indefinite future because they may be incorporated within ecological reserves or parks. The largest components of river systems, however, are unique and are subject to immediate change in connection with watershed development. This means that any significant changes in the main stem and major branches of the large rivers will forever deprive us of the opportunity to study the structure and functioning of these systems in their natural state. The priority of study should thus be highest for downstream rather than upstream portions of river systems, even though the hydrology and chemical properties of the systems will often be determined in their smaller branches.

The following subject areas should be included in studies of major rivers: water chemistry and nutrient processing, plankton, fishes, substrate fauna, key vertebrates other than fishes, and land–water interactions.

Water Chemistry and Nutrient Processing The chemical composition and discharge of major rivers are of great importance as indicators of the loss rates for organic and inorganic substances from tropical watersheds not subjected to human manipulation. The export rates of chemical substances are related to rates of weathering and to mechanisms of nutrient retention in the terrestrial portions of the watershed and therefore hold the key to any future interpretation of degree of change in these watersheds with respect to chemical export. Some background information on discharge already exists for most large rivers, but it is impossible to judge rates of chemical export without better studies of chemical composition of the main river stems and preferably at least of the major branches as well. In addition, chemical studies of this type would define the conditions to which the aquatic biota has become adapted before human alteration of the habitat. The load of suspended matter in these rivers should also be monitored as often as possible, and the values obtained should be correlated with deforestation and other major environmental changes. Hydrology itself is perhaps of even greater interest in these systems and should be studied in connection with chemical export. Water-storage mechanisms in large natural systems are not well understood, and study will be possible only until people interfere significantly.

Plankton The composition, abundance, and functioning of the plankton of large rivers in their natural state is essentially unstudied. Only the large tropical rivers offer such an opportunity; temperate rivers were grossly altered before they could be studied by modern methods. Because of the long residence time of water in such systems, the plankton can play an important role in the economy of the river and the processing of nutrients and organic substances. From what is now known, the taxonomic difficulties associated with studies of this type would be minimal; many of the important species are cosmopolitan or are known to temperate and tropical biologists dealing with other kinds of systems. The emphasis should therefore be placed on the plankton dynamics and their overall importance in the main stem of the river. Parallel or concurrent studies of plankton in the associated river wetlands will be mentioned later in connection with wetlands.

Fishes The taxonomy of fishes from the major tropical rivers mentioned above is being studied, but not rapidly enough (Böhlke et al., 1978, for South America). This activity should be encouraged in view of the possibility of widespread extinction. Data on life-history patterns, food webs, and behavior of the fishes are for the most part lacking (Lowe-McConnell, 1975). Major stocks of any of these fishes may be depleted to such an extent in the near future that it will be impossible to study these features of the fish community. This is especially true with respect to migrating fishes dependent on unimpeded access to upper regions of the river, which may easily be dammed.

Unless the ecology of fishes in major rivers is studied in the near future, we shall never know how these communities were organized, what their ranges of adaptation were, or how they exploited the original river habitat. Information of this type is of fundamental ecological importance. The fishes also represent an important human food resource that is insufficiently understood to be used with full effectiveness. In the same sense that the tropical rain forest might contain many species of trees whose products could be of great use, the fish communities of rivers may include members whose natural nutritional modes, behavior, or growth characteristics are useful in protein production or in the management of aquatic habitats.

Invertebrate Fauna The enormous variety of invertebrates in the major rivers presents fundamental taxonomic problems, and indeed the complexity is so great that the measurements may never be truly quantitative. For example, Junk (1973) found diverse assemblages as dense as 780,000 invertebrates/m^2 in the "floating meadows" of the middle Amazon. Only a few experts could be considered competent to deal systematically with the major invertebrate groups, which may be composed of thousands of species, many of them undescribed. A good catalog of systematists and references produced by Hurlbert (1977) gives specific information on the available expertise for much of South America. Important groups of great diversity are not infrequently the responsibility of only a handful of resident scientists; very limited help is available from the large museums of temperate regions. Moreover, many of the present experts are senior scientists with little time available for making identifications.

Why is it difficult to find newly trained scientists capable of, and interested in, meeting the basic taxonomic needs in freshwater and terrestrial groups of organisms in the tropics? The explanation probably is that most of these scientists have been trained in the temperate zone and are more interested in population dynamics than in systematics. This emphasis is in turn reflected in the training of tropical residents, who often take their advanced degrees in the temperate zone. Field work in the temperate zone is based on a long history of basic systematic and life-history study. This is not true in the tropics, and immediate support should therefore be given to training programs specifically focused on invertebrate taxonomy and life-history studies. A system of fellowships specifically keyed to these activities is needed, and persons receiving such fellowships should be carefully screened for dedication to these topics of study. Although it would take several years to launch such a program properly, it is essential that such a program be undertaken if we are to provide a taxonomic and life-history framework for studies more oriented toward processes and systems.

The ecology of invertebrates in the major rivers is a matter for immediate study. Many invertebrate communities will be especially sensitive to chemical changes in the rivers as they undergo development. The complexity and nutritional basis of the bottom-fauna communities, especially in connection with the processing of detritus, is of great theoretical interest to freshwater ecology and cannot be studied after the communities have been altered as a result of changes in the rivers. Knowledge of the bottom fauna is also essential for proper evaluation of data from fish communities.

Key Vertebrates Other Than Fishes   There is probably no need for immediate emphasis on some of the vertebrates associated with the riverine habitat, because they either do not play a key role in the ecology of the system or are not threatened by changes of the type that appear imminent. Certain vertebrates, however, may exert control over community structure and energy flow in natural aquatic systems in the tropics. For example, the crocodiles and their allies may have important effects on the distribution and abundance of fish populations; manatees and turtles are also of great interest. These key vertebrates need to be studied if an overall understanding of the system in its original condition is to be constructed before it is changed.

Land–Water Interactions Specific consideration must be given to the interactions between the flowing zone of the river, river wetlands (which will be discussed in greater detail later), and terrestrial systems. In a very large system not subject to significant human influence, mechanisms of nutrient transport and nutrient processing may exist that are unstudied because of the unavailability of such unaltered conditions in the temperate zone. A great increase in the fundamental understanding of large natural river systems could be achieved through study of these processes in the undeveloped large tropical rivers.

Smaller River Systems

Although detailed consideration could not be given to all the smaller rivers of the world, instances exist in which immediate major changes are expected in important rivers not connected with any of the major river systems that have already been mentioned. Two such rivers are the Musi in Sumatra and the Purari in Papua New Guinea. The large watersheds of these rivers, both of which are covered with wet forest, will probably be deforested in the near future. In addition, dams are contemplated for the Purari in the near future. The features of these rivers in their natural state might offer unique and important insights into rivers as natural systems. Unquestionably, other candidates for high priority will become evident among smaller rivers as exploitation of the moist forest proceeds.

LAKES

Compared with the number of lakes in the temperate zone, the tropics have few lakes, because in the tropics glaciation has played a role in lake formation only at the highest elevations. Because the study of temperate lakes has reached a quite sophisticated level, studies of tropical lakes will derive much of their value from comparison of tropical with temperate lakes. A large number of hypotheses related to the structure and functioning of lacustrine communities can be tested efficiently by study of tropical lacustrine systems because of major differences in important driving variables, such as temperature and light, that are impossible to manipulate experimentally in very large systems. For example, the relative importance of biotic, physical, and chemical factors in controlling the diversity and composition of the plankton continues to be unclear from temperate studies, and tropical studies will be of special value in resolving this issue.

In general, exploration of hypotheses by statistical means will be more feasible in tropical than in temperate systems because of the looser coupling between critical controlling variables, such as light and nutrient supply, in tropical lakes (Lewis, 1974, 1978). Tropical lakes are generally more productive than similar temperate ones (Talling, 1965, 1966; Lewis, 1974; Ganf, 1975) and may be more efficient in nutrient cycling. Primary production appears to translate into fish yield (Melack, 1976). Sharp contrasts with terrestrial systems, including apparent lack of marked diversity gradients in plankton at different latitudes, are intriguing. The reasons for this are obviously of general interest in ecology. Also of general interest and great potential importance to freshwater protein production is the low conversion efficiency of plankton primary productivity to plankton secondary productivity in the few tropical systems that have been studied in this way (Burgis, 1974; Lewis, 1979).

The preceding discussion has outlined several reasons for the continued and accelerated study of tropical lakes. Particular urgency exists for the study of those that may soon lose their original character, especially if they are of unusual scientific interest or regional importance. Three classes of lakes are of special interest for this reason and because of the factors just outlined.

Closed-Basin Lakes

Lakes of this type, which are most common in Africa (Beadle, 1974; Balek, 1977), lose water only by evaporation and consequently are often saline. They are especially vulnerable to human alteration because they do not flush. Closed-basin lakes that have been studied include Lake Nakuru (Kenya), Lake Valencia (Venezuela), and Lake Chad (Chad). These lakes tend to be extremely productive and rich in plant biomass. The more saline ones have very simple communities and short food chains. Such lakes may be ideal for study of certain ecological principles, including the factors that set absolute upper limits on net primary production in aquatic systems and the factors that lead to extreme dominance by one or a few species.

Lakes with High Levels of Endemism

Flocks of endemic fish species are found in several tropical lakes, especially those of great age (Brooks, 1950; Fryer and Iles, 1972). The urgency with regard to endemism derives from the vulnerability of species flocks to disruption by means of species introductions. Introductions are commonplace and probably cannot be stopped.

The East African Rift Valley lakes contain the most spectacular fish species flocks. One of these is Lake Malawi, which is important not only because it has the largest number of endemic species (more than 200 species) but also because it urgently requires study for other reasons (see below). Smaller species flocks in other lakes around the world should also be studied, especially since their endemics are rapidly disappearing. The scientific merit of studies of species flocks is related to the insight that may be obtained into the rates and mechanisms of evolution.

Lakes Likely To Be Altered by Development

In this category Lake Malawi, Lake Titicaca, the Sunda lakes of Sumatra and insular Southeast Asia generally, and Lake Maracaibo (Venezuela) are important.

Because of its great depth and its chemical purity, Lake Malawi is of special interest. The lake is very old and extremely deep. Long-term changes in climate have undoubtedly influenced the present balance between physical processes of layering and mixing and the biological and chemical processes of nutrient uptake and mineralization. Analysis of chemical profiles, chemical input, and biological nutrient-cycling processes under these conditions may be revealing. Onset of significant chemical alterations because of pollution or other mechanisms would vastly reduce the potential of such an analysis.

Lake Titicaca is the only very large (8,100 km^2) perennially cold lake in the tropics, and current studies (Richerson *et al.*, 1977; Orstom, personal communication, 1979) should be intensified. The unusual combination of conditions in this lake has enormous comparative value with respect to the biological and chemical phenomena that occur in lakes generally; moreover, the lake has a large number of endemic species. The lake has been altered greatly at the upper trophic levels by introduction of exotic fishes. There are many smaller lakes in the Andes

Tropical Aquatic Systems

of Peru and Bolivia that are still unaltered but which resemble Lake Titicaca in many characteristics. They might certainly be studied on a comparative basic.

The lakes of insular Southeast Asia, including the very large Lake Toba in Sumatra and a number of smaller lakes, are of special interest. Many of these were studied in the 1920's by the German Sunda Expedition (Ruttner, 1952). The watersheds of these lakes are being changed very rapidly by deforestation and other kinds of development. The comprehensive data base gathered by the Sunda Expedition provides favorable circumstances for selected additional studies. It is important to obtain a second set of analyses before development but well separated in time from the first to provide a basis for judgment of natural rates of change in tropical lakes over extended time periods (Green et al., 1976).

Lake Maracaibo is of interest because of its great size and its connection with the sea. Petroleum exploitation and other development seem certain to alter this lake drastically in the near future, and some changes are already evident.

WETLANDS

Freshwater wetlands have been little studied even in the temperate zone (Good et al., 1978). The habitats that fall into this classification have a wide range of physical, chemical, and biological characteristics.

Some wetlands are among the most productive of freshwater systems, and they may serve as massive processing units for organic matter and nutrients brought in from other systems. The humid tropics contain many wetlands, including a few of enormous size. The total area of tropical swamps is thought to be about 340,000 km^2, about one quarter of which is seasonal (Balek, 1977). The large wetlands are likely to be unique in many respects and are extremely vulnerable to destruction by drainage or alteration of associated riverine systems. Accordingly, they are in need of immediate study, especially since they and the swamps along the deltas of rivers are so important as sites for concentrated and productive rice agriculture.

Large Swamps

Large swamps that stand in imminent danger of alteration and thus have high priority for study include the Sudd (Sudan), the Benguelu swamps of Zambia, and the Okavango swamp of Botswana; the swamps along the Sepik and Fly rivers of Papua New Guinea, as well as those of Territoria Amapá of Brazil and of Bení Department of Bolivia; and the Pantanal of Mato Grosso in Brazil, which is dry for part of the year.

The highest priority is accorded to the Sudd, a massive set of swamps of unique character maintained by the Nile (Rzóska, 1976). The permanent swamp has an extent of at least 10,000 km² and expands to 70,000 km² or more seasonally. Major drainage projects have already been initiated in this huge swamp. The 173,000 km² of Pantanal in Mato Grosso, Brazil, will probably be drained in the near future. The other large wetlands will probably not be significantly altered quite so quickly, but they should receive relatively high priority because of their great vulnerability. Additional swamps too numerous to mention individually will quickly move up in priority if the associated forests are cleared. For example, Balek (1977) lists a dozen swamps of over a thousand square kilometers' area in Africa alone.

The swamps named here are likely to have distinctive species-composition and energy-flow characteristics. The exact plan for research will vary according to the characteristics to be studied. In general, however, research should proceed as quickly as possible toward characterization of these wetlands as ecological systems. Intensive study of individual components, including the dominant plants, will probably be possible for much longer than the study of ecosystem processes.

The following topics are of interest as foci for research in large permanent swamps because of their value in interpreting the functioning of these and other freshwater systems.

Factors Regulating Energy Flow   Parts of some of these wetlands are probably among the most productive of aquatic ecosystems. The mechanisms by which such high primary production is sustained and the characteristics of the dominant organisms are obviously of basic interest. In addition to their high primary production, energy flow through the heterotrophic components will probably be of especial interest in many cases. The role of swamps as intermediary metabolic units is important in nutrient cycling and has enoumous practical importance with regard to natural purification processes in freshwater systems. The large permanent swamps offer unique opportunities to study these processes on a large scale.

Ecology of Key Species   Wetlands are sometimes dominated by a few species. The best example of such dominance is papyrus in African wetlands. In cases in which species domination is found, it would be profitable to focus attention on the ecology and adaptive characteristics of the dominant plants but without losing the context of the system within which they normally grow.

Special Adaptations Permanent wetlands contain many species with special adaptations to seasonal fluctuations in water level and the special chemical characteristics of the wetlands habitat (Beadle, 1974). Elucidation of these adaptations, whether they are morphological or connected with life-history phenomena, is of fundamental interest. Important topics include special colonizing ability, which may be essential for many of the dominant organisms, and adaptations for seasonally unfavorable conditions, such as anoxia or desiccation.

Major Riverine Wetlands

The distinction between this and the preceding category of wetlands is not sharp, but we refer here specifically to wetlands that are actually incorporated into river flow at times of high water but then become isolated at low water. Each of the large rivers mentioned earlier is paralleled by extensive associated wetlands that are flooded seasonally at high water and dry out substantially during periods of low discharge.

These riverine wetlands must ultimately be evaluated in conjunction with the rivers themselves. The methods of study will be quite different from those used for the river stem, because the river wetlands are nonflowing environments most of the year. The research topics relevant to these wetlands are the same as those listed for rivers, but greater emphasis should be placed on the role of wetlands as nutrient traps, as sources of organic matter and energy, and as refugia for organisms with seasonal cycles geared to the annual inundation of the areas. A significant research effort dealing with these topics is in progress in the Amazon Basin. It is under the supervision of the Instituto Nacional de Pesquisas da Amazônia (INPA). The studies are centered at the permanent INPA field station, Manaus, but they should probably be expanded and distributed to other major rivers while there is still time to study the wetlands in their original state.

Highest priority should be given to the study of wetlands in the lower reaches of the Amazon and Orinoco rivers and to their *várzea*. We would then assign secondary priority to the wetlands of the Zaire and to the extensive wetlands associated with the upper reaches of the Xingu.

TABLE 1 Suggested Time Scale for High-Priority Research in Tropical Aquatic Systems

Systems (Studies) in Priority Order	Time Within Which Study Must Be Initiated (Years)		
	5	10	15
Rivers			
Very large rivers	Amazon, Orinoco	Zaire	Mekong, Nile
Smaller rivers of special importance	Purari	Musi	As appropriate
Lakes			
Closed-basin lakes	Valencia	Malawi	Nakuru
Important lakes subject to major change	Valencia	Titicaca	Lakes of insular Southeast Asia
Lakes containing species flocks	Maracaibo	Malawi Titicaca	Other African lakes Other small lakes
Wetlands			
Large swamps	Sudd Pantanal Mato Grosso Okavango	Territorio Amapá	Bengweulu Provincia Beni
Riverine wetlands	Amazon *várzea* Orinoco Delta and backwaters	Zaire Xingu	
Other wetlands		Southeast Asian peat swamps	As appropriate
Precipitation		Mekong, Amazon, Orinoco	Zaire

Peat Swamps

The lowlands tropics of Southeast Asia contain some 30,000 km^2 of peat bogs. These occur as units of up to 500 km^2 whose sole source of water is rain or groundwater flow. These tropical bogs have been insufficiently studied, will be lost or changed during deforestation, and are of special interest for comparison with temperate bogs. Like temperate bogs, these tropical bogs are likely to provide especially valuable paleoclimatic and paleobotanical information. Nutrient cycling is of great interest in these areas because of their extreme oligotrophy.

PRECIPITATION

The entire hydrologic cycle is relevant to the characteristics of freshwater environments. Those stages of the cycle in which water is incorporated in soil or plants may be most easily studied by those who deal with terrestrial ecosystems. Precipitation amounts and chemistry might logically be studied both by those concerned with terrestrial habitats and by those concerned with aquatic ones. For this reason we include precipitation as a distinct topic for research in connection with the study of aquatic habitat in the tropics. Such studies could be done by a group of scientists specifically interested in precipitation or by scientists primarily interested in terrestrial or aquatic habitats as a support for studies of such habitats. In addition to the usual kinds of studies, monitoring is needed of the total chemical input to the large watersheds from the atmosphere, preferably by means of continuous collection of dry and wet precipitation followed by chemical analysis for components of particular interest with respect to the nutrition of plants and chemical buffering of aqueous mixtures. For studies of precipitation, a broadly spread network such as that described by Salati *et al.* (1979) for the Amazon would provide useful information.

SUMMARY OF PRIORITIES

Table 1 summarizes our priorities for studies and indicates the time within which studies must be initiated to provide meaningful information before alteration of the ecosystem in question. An attempt has been made to spread the research topics sufficiently to allow reasonable possibilities for funding of the projects. The table states only the most pressing priorities. It is not an overall statement of tropical freshwater research needs.

If the suggestions given in column 1 of Table 1 were followed, the annual cost of investigations would probably be between $3 million and $5 million per year. Addition of the research topics listed in the second column of the table would raise the cost to at least triple this amount. Many of the studies could be of restricted duration (e.g., 5–10 years), permitting phasing of the projects to prevent unrealistic escalation of costs. In fact, the scope of the research plan is such that intensive projects would almost certainly have to be of restricted duration in all but a few cases, although locally supported monitoring should continue indefinitely following intensive study. Careful planning and staging of the projects would be essential to the completion of any major portion of the work. Investments greater than those implied in Table 1 would probably be unrealistic because of current limits on personnel qualified to do the work.

References

Adis, J. 1977. Programa mínimo para analisis de ecosistemas: artrópodos terrestres em florestas inundavais da Amazônia central. Acta Amazonica 7(2):223–229.
Adis, J. 1979. Vergleichende oekologische Studien an der terrestrischen Arthropodenfauna vergleichende zentralamazonischer ueberschwemmungswaelder. Unpublished Ph.D. dissertation, University of Ulm, Federal Republic of Germany.
Alvim, P. de T., and T. T. Kozlowski. 1977. Ecophysiology of Tropical Crops. Academic Press, New York.
Anonymous. 1974a. Trends, priorities, and needs in systematic and evolutionary biology. Syst. Zool. 23:416–439.
Anonymous. 1974b. Programme on Man and the Biosphere (MAB). International working group on Project 1: Ecological effects of increasing human activities on tropical and subtropical forest ecosystems. MAB Rep. Ser. No. 16. UNESCO, Paris.
Anonymous. 1976. International co-ordinating council of the Programme on Man and the Biosphere (MAB), Fourth Session. MAB Rep. Ser. No. 38. UNESCO, Paris.
ABRC (U.K. Advisory Board for the Research Councils). 1979. Taxonomy in Britain. H. M. Stationery Office, London.
Arnold, J. E. M., and J. Jongma. 1978. Fuelwood and charcoal in developing countries. Unasylva 29:2–9.
Arroyo, M. T. K. 1976. Geitonogamy in animal pollinated tropical angiosperms. A stimulus for the evolution of self-incompatibility. Taxon 25:543–548.
Ashton, P. S. 1969. Speciation among tropical forest trees: some deductions in the light of recent evidence. Biol. J. Linn. Soc. 1:155–196.
Ashton, P. S. 1970. The biological significance of complexity in lowland tropical rain forest. J. Indian Bot. Soc. 50A:530–537.
Ashton, P. S. 1976a. An approach to the study of breeding systems, population structure and taxonomy of tropical trees. Pages 35–42 *in* J. Burley and B. T. Styles, eds. Tropical Forest Trees: Variation, Breeding Systems, and Conservation. Linnaean Society Symposium Series 2. Academic Press, London.

Ashton, P. S. 1976b. Factors affecting the development and conservation of tree genetic resources in South-east Asia. Pages 189–198 in J. Burley and P. T. Styles, eds. Tropical Forest Trees: Variation, Breeding Systems, and Conservation. Linnaean Society Symposium Series 2. Academic Press, London.

Ashton, P. S., and E. F. Brünig. 1975. The location of tropical moist forest in relation to environmental factors in its relevance to land use planning. Mitt. Bundesforschungsanst. Forst. Holzwitsch. (Hamburg) 109:59–86.

Ashton, P. S., M. J. Hopkins, L. J. Webb, W. T. Williams, and J. Palmer. 1978. Description, functioning and evolution of tropical forest ecosystems. The natural forest: plant biology, regeneration and tree growth. Pages 180–215 in UNESCO, UNEP, FAO Natural Resources Research XIV. Tropical Forest Ecosystems: A State-of-Knowledge Report. UNESCO, Paris, France.

Aubréville, A. 1961. Etude écologique des principales formations végétales du Brésil et contribution à la connaissance des forêts de l'Amazonie brésilienne. Centre Technique Forestier Tropicale, Nogent-sur-Marne, France. 268 pp.

Auchter, R. J. 1978. 1978 Summation. Pages 1–22 in Proceedings of Conference on Improved Utilization of Tropical Forests. USDA-Forest Service, Forest Products Laboratory, Madison, Wis.

Ayensu, E. S. 1978. Conserving the African environment. Africa Report 3 (May–June): 14.

Baker, H. G. 1975. Sugar concentrations in nectars from hummingbird flowers. Biotropica 7:37–41.

Balek, J. 1977. Hydrology and Water Resources in Tropical Africa. Elsevier, Amsterdam, The Netherlands. 208 pp.

Bawa, K. S. 1973. Chromosome numbers of tree species of a lowland tropical community. J. Arnold Arb. Harv. Univ. 54:422–434.

Bawa, K. S. 1974. Breeding systems of tree species of a lowland tropical community. Evolution 28:85–92.

Bawa, K. S. 1976. Breeding of tropical hardwoods: an evaluation of underlying bases, current status and future prospects. Pages 43–59 in J. Burley and B. T. Styles, eds. Tropical Forest Trees: Variation, Breeding Systems, and Conservation. Linnaean Society Symposium Series 2. Academic Press, London.

Bawa, K. S., and P. A. Opler. 1975. Dioecism in tropical forest trees. Evolution 29:167–179.

Beadle, L. C. 1974. The Inland Waters of Tropical Africa: An Introduction to Tropical Limnology. Longman, Inc., New York.

Bergman, R. W. 1975. Shipibo Subsistence in the Upper Amazon. Unpublished Ph.D. dissertation, University of Wisconsin, Madison.

Bernardi, L. 1974. Problèmes de conservation de la nature dans les îles de l'Ocean Indien. 1. Meditation à propos de Madagascar. Saussurea 5:37–47.

Böhlke, J. E., S. H. Weitzman, and N. A. Menezes. 1978. Estado atual da sistemática dos peixes de água doce da América do Sul. Acta Amazonica 8(4):657–677.

Bolin, B. 1977. Changes of land biota and their importance for the carbon cycle. Science 196:613–615.

Bormann, F. H., and G. E. Likens. 1979. Pattern and Process in a Forested Ecosystem. Springer-Verlag, New York. 253 pp.

Brenan, J. P. M. 1978. Some aspects of the phytogeography of tropical Africa. Ann. Mo. Bot. Gard. 65:437–478.

Bradfield, R. 1972. Maximizing food production through multiple cropping systems centered on rice. Pages 143–163 in Rice Science and Man. International Rice Research Institute, Los Banos, the Philippines.

References

Broecker, W. S., T. Takahashi, H. J. Simpson, and T. H. Peng. 1979. Fate of fossil fuel carbon dioxide and the global carbon budget. Science 206:409–418.
Brooks, J. L. 1950. Speciation in ancient lakes. Q. Rev. Biol. 25:30–60, 131–176.
Brown, L. 1976. World Population Trends: Signs of Hope, Signs of Stress. Worldwatch Paper 8. Worldwatch Institute, Washington, D.C. 40 pp.
Brünig, E. F. 1977. Transactions of the International MAB–IFURO Workshop on Tropical Rainforest Ecosystems Research, Hamburg-Reinbek, 12.-17.5.1977. Chair of World Forestry, Hamburg-Reinbek, Special Report No. 1. 353 pp.
Bryson, R. A., and T. J. Murray. 1977. Climates of Hunger. University of Wisconsin Press, Madison.
Budowski, G. 1974. Forest plantations and nature conservation. IUCN Bull. 5:25–26.
Budowski, G. 1976. Why save tropical forests? Some arguments for campaigning conservationists. Amazoniana 4:529–538.
Budowski, G. 1977. A strategy for saving wild plants: Experience from Central America. Pages 368–373 *in* G. T. Prance and E. S. Elias, eds. Extinction is Forever. New York Botanical Garden, Bronx, New York.
Budyko, M. I. 1974. Climate and Life. Academic Press, New York.
Burgis, M. J. 1974. Revised estimates for the biomass and production of zooplankton in Lake George, Uganda. Freshwater Biol. 4:535–541.
Burley, J., and B. T. Styles, eds. 1976. Tropical Forest Trees: Variation, Breeding, and Conservation. Linnaean Symposium Series 2. Academic Press, New York.
Calvin, M. 1976. Photosynthesis as a resource for energy and materials. Photochem. Photobiol. 23: 425:444.
Carson, H. L., and K. Y. Kaneshiro. 1976. *Drosophila* of Hawaii. Systematics and ecological genetics. Ann. Rev. Ecol. Syst. 7:311–345.
Casey, T. L. C., and J. D. Jacobi. 1974. A new genus and species of bird from the island of Maui, Hawaii (Passeriformes; Drepanididae). Occ. Pap. Bernice Pauahi Bishop Mus. 24:215–226.
Choudhury, B., and G. Kukla. 1979. Impact of CO_2 on cooling of snow and water surfaces. Nature 280:668–671.
Connell, J. H. 1978. Diversity in tropical rain forests and coral reefs. Science 199:1302–1310.
Croat, T. B. 1979. Flora of Barro Colorado Island. Stanford University Press, Stanford, Calif.
Crutzen, P. J., L. E. Heidt, J. P. Krasnec, W. H. Pollock, and W. Seiler. 1979. Biomass burning as a source of atmospheric gases CO, H_2, N_2O NO, CH_3Cl and CO_2. Nature 282:253–256.
Dahlberg, K. A. 1979. Beyond the Green Revolution. The Ecology and Politics of Global Agricultural Development. Plenum Press, New York.
d'Arge, R. 1979. Climate and economic activity. Paper presented at the U.N. Conference on World Climate, Geneva, February 1979.
Dasmann, R. F., J. P. Milton, and P. H. Freeman. 1973. Ecological Principles for Economic Development. John Wiley & Sons, New York.
Dawkins, H. C. 1958. The management of natural tropical highforest with special reference to Uganda. Imperial Forestry Institute Paper No. 34. Imperial Forestry Institute, Oxford.
Deevey, E. S., D. S. Rice, P. M. Rice, H. H. Vaughan, M. Brenner, and M. S. Flannery. 1979. Mayan urbanism: impact on a tropical karst environment. Science 206:298–306.
Diamond, J. M. 1972. Biogeographic kinetics: estimation of relation times for avifaunas of southwest Pacific islands. Proc. Natl. Acad. Sci. U.S.A. 69:3199–3203.

Diamond, J. M. 1975. The island dilemna: lessons of modern biogeographic studies for the design of natural preserves. Biol. Conserv. 7:129–146.

Diamond, J. M. 1976. Island biogeography and conservation: strategy and limitations. Science 193:1027–1032.

Di Castri, F., and M. Hadley. 1978. Ecological approaches to land resources in the tropics: some case studies from the Man and Biosphere (MAB) Programme. Pages 569–586 in J. S. Singh and B. Gopal, eds. Glimpses of Ecology. (Professor R. Misra Commemoration Volume). International Scientific Publishers, Jaipur, India.

Di Castri, F., and M. Hadley. In press. A typology of scientific bottlenecks to natural resources development. GeoJournal 3(6).

Dodson, C. H., and A. H. Gentry. 1978. Flora of the Rio Palenque Science Center, Los Rios Province, Ecuador. Selbeyana 4:i–xxix, 1–628.

Duellman, W. E. 1978. The biology of an equatorial herpetofauna in Amazonian Ecuador. Univ. Kans. Mus. Nat. Hist. Misc. Publ. 65:1–352.

Eckholm, E. 1976. Losing Ground. Environmental Stress and World Food Prospects. W. H. Norton & Co., Inc., New York.

Eckholm, E. 1979a. The Dispossessed of the Earth: Land Reform and Sustainable Development. Worldwatch Paper 30. The Worldwatch Institute, Washington, D.C. 48 pp.

Eckholm, E. 1979b. Planting for the Future: Forestry for Human Needs. Worldwatch Paper 26. The Worldwatch Institute, Washington, D.C. 64 pp.

Ehrenfield, D. W. 1972. Conserving Life on Earth. University Press, New York.

Ehrenfeld, D. W. 1976. The conservation of non-resources. Am. Sci. 64:648–656.

Ehrlich, P. R., A. H. Ehrlich, and J. P. Holdren. 1977. Ecoscience: Population, Resources, Environment. W. H. Freeman, San Francisco, Calif.

ESRC (European Science Research Council). 1977. Taxonomy in Europe. Committee of European Science Research Councils, ESRC Review No. 13. European Science Foundation, Strasbourg, France.

Farnworth, E., and F. Golley, eds. 1974. Fragile Ecosystems. Evaluation of Research and Applications in the Neotropics. Springer-Verlag, Berlin, Heidelberg, and New York.

Finn, J. T. 1976. Measures of ecosystem structure and function derived from analysis of flows. J. Theor. Biol. 56:363–380.

Finn, J. T. 1978. Cycling index: a general definition for cycling in compartment models. Pages 138–164 in D. C. Adriano and I. L. Brisbin, eds. Environmental Chemistry and Cycling Processes. Technical Information Center, U.S. Department of Energy, Washington, D.C.

Fitzpatrick, J. W., J. J. Terborgh, and D. E. Willard. 1977. A new species of wood-wren from Peru. Auk 94(2):195.

Flenley, J. R. 1979. The Equatorial Rain Forest: A Geological History. Butterworths, London.

Flint, O. S., Jr. 1971. Studies of neotropical caddisflies. XII. Rhyacophilidae, Glossomatidae, Philipotamidae, and Psychomyiidae from the Amazon Basin (Trichoptera). Amazonia III (1):1–67.

Fournier, L. A. 1969. Observaciones preliminares sobre la variación altitudinal en el numero de familias de arboles y de arbustos en la vertiente del Pacífico de Costa Rica. Turrialba 19:548–552.

Fox, J. E. D. 1976. Constraints on the natural regeneration of tropical moist forest. For. Ecol. Manage., 1:37–65.

Frankie, G. W., H. G. Baker, and P. A. Opler. 1974a. Comparative phenological studies of trees in tropical wet and dry forests in the lowlands of Costa Rica. J. Ecol. 62:881–919.

References

Frankie, G. W., H. G. Baker, and P. A. Opler. 1974b. Tropical plant phenology: applications for studies in community ecology. Pages 287–296 *in* H. Lieth, ed. Phenology and Seasonality Modeling. Springer-Verlag, New York.

Frankel, O. H. 1977. Natural variation and its conservation. Pages 21–44 *in* A. Muhammed, R. Aksel, and R. C. von Borstel, eds. Genetic Diversity in Plants. Plenum Press, New York.

Frankel, O. H., and E. Bennett, eds. 1970. Genetic Resources in Plants—Their Exploration and Conservation. IBP Handbook No. 11. F. A. Davis Co., Philadelphia, Pa.

Frankel, O. H., and J. G. Hawkes, eds. 1975. Crop Genetic Resources for Today and Tomorrow. International Biological Programme 2. Cambridge University Press, London, New York, Melbourne.

Friedman, I. 1977. The Amazon Basin, another Sahel? Science 197:7.

Fryer, G., and T. D. Iles. 1972. The Cichlid Fishes of the Great Lakes of Africa: Their Biology and Evolution. T. F. H. Publications, Neptune City, N.J.

Ganf, G. G. 1975. Photosynthetic production and irradiance–photosynthesis relationships of the phytoplankton from a shallow equatorial lake (Lake George, Uganda). Oecologia (Berlin) 18:165–183.

Gentry, A. H. 1978. Floristic knowledge and needs in Pacific tropical America. Brittonia 30:134–153.

Gliessman, S. R. 1979. Avances en el estudio de las bases ecológicas de la producción en algunos agroecosistemas de Tabasco, México. Paper presented at II Seminario de Analisis de los Agroecosistemas de México, 16–21 de julio, Chapingo, México.

Goldsmith, E. 1980. World Ecological Areas Program (WEAP). The Ecologist 10(1/2):1–4.

Gómez-Pompa, A. 1967. Some problems of tropical plant ecology. J. Arnold Arbor. Harv. Univ. 48:104–121.

Gómez-Pompa, A. 1972. Posible papel de la vegetación secundaria en la evolución de la flora tropical. Biotropica 3:125–135.

Gómez-Pompa, A. 1978. Ecología de la Vegetación de Veracruz. Compañia Editorial Continental, S.A., México, D.F.

Gómez-Pompa, A., C. Vásquez-Yanes, and S. Guevara. 1972. The tropical rain forest: a nonrenewable resource. Science 177:762–765.

Gómez-Pompa, A., and A. Butando C. 1975. Index of Current Tropical Ecology Research. Vol. 1. Consejo Nacional de Ciencia y Tecnología. Instituto Nacional de Investigaciones sobre Recursos Bióticos, México, D.F.

Gómez-Pompa, A., S. de Amo, R. C. Vásquez-Yanes, and A. Butando C., eds. 1976. Investigaciones sobre la Regeneración de las Selvas Altas en Veracruz, México. Compañia Editorial Continental, México, D.F.

Gómez-Pompa, A., and A. Butanda C. 1977. Index of Current Tropical Ecology Research. Vol. 2. Consejo Nacional de Ciencia y Tecnología Programa Nacional Indicativo de Ecología. Instituto Nacional de Investigaciones sobre Recursos Bióticos, Xalapa, Veracruz, México.

Good, R. E., D. F. Whigham, and R. L. Simpson, eds. 1978. Freshwater Wetlands: Ecological Processes and Management Potential. Academic Press, New York.

Goodland, R. J. A., and H. S. Irwin. 1975. Amazon Jungle: Green Hell to Red Desert? Elsevier Scientific Publishing Company, Amsterdam, the Netherlands. 155 pp.

Goodland, R. J., and H. S. Irwin. 1977. Amazonian forest and cerrado: development and environmental conservation. Pages 214–233 in G. T. Prance and T. S. Elias, eds. Extinction is Forever. The Status of Threatened and Endangered Plants of the Americas. New York Botanical Garden, Bronx, New York.

Gray, B. 1974. The economics and planning of research into tropical forest insect pests. Pest Articles and News Summaries 20(1):1–10.

Green, J., S. A. Corbet, E. Watts, and O. B. Lan. 1976. Ecological studies on Indonesian lakes. Overturn and restratification of Ranu Lamongan. J. Zool. (Lond.) 180:315–354.

Haantjens, H. A. 1975. Papua New Guinea: an example of conservation opportunities in the humid tropics. Search 6:477–484.

Hallé, F., R. A. A. Oldeman, and P. B. Tomlinson. 1978. Tropical Trees and Forests. Springer-Verlag, Berlin. 441 pp.

Hanlon, J. 1979. When the scientist meets the medicine man. Nature 279:284–285.

Harney, T. 1979. Museum collections help identify a deadly mosquito. Smithson. Inst. Res. Rep. 27:3.

Hartshorn, G. S. 1978. Tree falls and tropical forest dynamics. Pages 617–638 in P. B. Tomlinson and M. H. Zimmermann, eds. Tropical Trees as Living Systems. Cambridge University Press, Cambridge, U.K.

Hawkes, J. G. 1977a. The importance of wild germplasm in plant breeding. Euphytica 26:615–621.

Hawkes, J. G. 1977b. Plant gene pools—an essential resource for the future. J. R. Soc. Arts 125:224–235.

Hawkes, J. G., ed. 1979. Conservation and Agriculture. Duckworth, London.

Heckadon, S. P. 1979. Dinámica social de la cultura del potrero en Panamá. J. Trop. Ecol. (Veranasi, India) 19(2).

Herrera, R., T. Merida, N. Stark, and C. F. Jordan. 1978a. Direct phosphorus transfer from leaf litter to roots. Naturwissenschaften 65:208–209.

Herrera, R., C. F. Jordan, H. Klinge, and E. Medina. 1978b. Amazon ecosystems. Their structure and functioning with particular emphasis on nutrients. Interciencia 3:223–232.

Heslop-Harrison, J. 1973. The plant kingdom: an exhaustible resource? Trans. Bot. Soc. Edinburgh 42:1–15.

Heslop-Harrison, J. 1974. Genetic resource conservation: the end and the means. J. R. Soc. Arts 122:157–169.

Hubbell, S. P. 1979. Tree dispersion, abundance, and diversity in a tropical dry forest. Science 203:1299–1309.

Humbert, J. 1927. Destruction d'une flore insulaire par le feu. Principaux aspects de la végétation à Madagascar. Mém. l'Acad. Malgache 5:1–80.

Hurlbert, S. H., ed. 1977. Biota Acuática de Sudamerica Austral. San Diego State University, San Diego, Calif. 342 pp.

Hutchinson, G. E. 1954. Pages 371–433 in E. P. Kuiper, ed. Biogeochemistry of the Terrestrial Atmosphere. Solar Systems, Vol II. University of Chicago Press, Chicago, Ill.

Hutchinson, I. D. 1979. A practical guide to liberation thinning—a type of silvicultural treatment to be applied in recently logged mixed Dipterocarp forests of Sarawak. UNDP/FAO/MAL/76/008, Sarawak Forest Department, Kuching, Malaysia.

Innis, D. Q. 1972. The efficiency of Jamaican peasant land use. Can. Geogr. 5(2):19–23.

International Food Policy Research Institute. 1977. Food Needs of Developing Countries: Projections of Prodution and Consumption to 1990. Res. Rep. No. 3. International Food Policy Research Institute, Washington D.C. 157 pp.

Irion, G. 1978. Soil infertility in the Amazonian rain forest. Naturwissenschaften 65:515–519.

Iltis, H. H. 1972. The extinction of species in the destruction of ecosystems. Am. Biol. Teach. 34:201–205, 221.

Iltis, H. H. 1974. Freezing the genetics landscape—the preservation of diversity in cultivated plants as an urgent social responsibility of the plant geneticist and plant taxonomist. Maize Genet. Coop. Newsl. 48:199–200.

References

Irwin, H. S. 1977. Coming to terms with the rain forest. Garden 1(2):28–33.
Jacobs, M. 1977. Gardens, species, universities. Pages 211–217 *in* B. C. Stone, ed. The Role and Goals of Tropical Botanic Gardens. Rimba Ilmu Universiti Malaya, Kuala Lumpur, Malaysia.
Jacobs, M. 1979. Conservation. Flora Malesiana Bull. 32:3227–3238.
Janzen, D. H. 1967. Interaction of the bull's horn acacia (*Acacia cornigera* L.) with an ant inhabitant (*Pseudomyrmex ferruginea* F. Smith) in eastern Mexico. Univ. Kans. Bull. 47:315–558.
Janzen, D. H. 1972. The uncertain future of the tropics. Nat. Hist. 81:80–89.
Janzen, D. H. 1973a. Tropical agroecosystems. Science 182:1212–1219.
Janzen, D. H. 1973b. Sweep samples of tropical foliage insects: description of study sites, with data on species abundance and size distributions. Ecology 54:659–686.
Janzen, D. H. 1973c. Sweep samples of tropical foliage insects: effects of seasons, vegetation types, elevation, time of day and insularity. Ecology 54:687:708.
Janzen, D. H. 1974a. The deflowering of Central America. Nat. Hist. 83:48–53.
Janzen, D. H. 1974b. Tropical blackwater rivers, animals, and mast fruiting by the Dipterocarpaceae. Biotropica 6:69–103.
Janzen, D. H. 1976. Why bamboos wait so long to flower. Ann. Rev. Ecol. Syst. 7:347–391.
Janzen, D. H. 1977a. Addtional land at what price? Responsible use of the tropics in a food-production confrontation. Proc. Am. Phytopath. Soc. 3:35–39.
Janzen, D. H. 1977b. The impact of tropical studies on ecology. Pages 159–187 *in* The Changing Scenes in Natural Sciences, 1776–1976. Acad. Nat. Sci. Spec. Publ. 12. Philadelphia, Pa.
Jordan, C. F. 1979. Stem flow and nutrient transfer in a tropical rain forest. Oikos 31:257–263.
Jordan, C. F., and E. Medina. 1978. Ecosystem research in the tropics. Ann. Mo. Bot. Gard. 64:737–745.
Jordan, C. F., and N. Stark. 1978. Retención de nutrientes en la estera de raices de un bosque pluvial Amazonico. Acta Cient. Venez. 29:263–267.
Jordan, C. F., and C. Uhl. 1978. Biomass of a "tierra firme" forest of the Amazon Basin. Oecol. Plant. 13(4):387–400.
Jordan, C. F., R. L. Todd, and G. Escalante. 1979a. Nitrogen conservation in a tropical rain forest. Oecologia (Berlin) 39:123–128.
Jordan, C. F., F. Golley, J. D. Hall, and J. Hall. In press. Nutrient scavenging of rainfall by the canopy of an Amazonian rain forest. Biotropica.
Junk, W. J. 1973. Investigations on the ecology and production-biology of the "floating meadows" (Paspalo-Echinochloetum) on the middle Amazon. Part II: The aquatic fauna in the root zone of floating vegetation. Amazoniana 4(1):9–102.
Kato, R., Y. Tadaki, and H. Ogawa. 1978. Plant biomass and growth increment studies in Pasoh Forest. Malay. Nat. J. 30(12):211–224.
Kavanaugh, M. 1979. The world's vanishing primates. Nature 277:432–434.
Kemp, R. H. 1978. Exploration, utilization and conservation of genetic resources. Unasylva 30:10–16.
Kemp, R. H., L. Roche, and R. L. Willan. 1976. Current activities and problems in the exploration and conservation of tropical forest gene resources. Pages 223–233 *in* J. Burley and B. T. Styles, eds. Tropical Forest Trees: Variation, Breeding Systems, and Conservation. Linnaean Society Symposium Series 2. Academic Press, London.
Kira, T. 1978. Community architecture and organic matter dynamics in tropical lowland rain forest of Southeast Asia with special reference to Pasoh Forest, West Malaysia. Pages 561–590 *in* Tropical Trees as Living Systems. Cambridge University Press, Cambridge, U.K.

Klinge, H., E. Medina, and R. Herrera. 1977. Studies on the ecology of Amazon caatinga forest in southern Venezuela. Acta Cient. Venez. 28:270–276.

Knutson, L. 1978. Uses and user community of entomological collections. Entomol. Scand. 8(2):155–160.

Lachner, E. A., J. W. Atz, G. W. Barlow, B. B. Collette, R. J. Lavenberg, C. R. Robins, and R. J. Schultz. 1976. A national plan for icthyology. Copeia 1976:618–625.

Lamb, K. P. 1974. Economic Entomology in the Tropics. Academic Press, London. 195 pp.

Lamotte, M., and F. Bourlière, eds. 1978. Problèmes d'Écologie: Structure et Fonctionnement des Écosystèmes Terrestres. Masson, Paris. 345 pp.

Lemée, G., C. Huttel, and F. Bernhard-Reversat. 1975. Recherches sur l'écosystème de la forêt subéquatoriale de Basse Côte d'Ivoire. La Terre et la Vie 29:169–264.

Lettau, H. et al. 1979. Amazonia hydrologic cycle and the role of atmosphere recycling in assessing deforestation effects. Mon. Weather Rev. 107(3):227–238.

Leroy, J. F. 1978. Composition, origin, and affinities of the Madagascan vascular flora. Ann. Mo. Bot. Gard. 65:535–589.

Letouzey, R. 1968. Etude Phtogéographique du Cameroun. Lechevalier, Paris. 511 pp.

Lewis, W. M., Jr. 1974. Primary production in the plankton community of a tropical lake. Ecol. Monogr. 44(4):377–409.

Lewis, W. M., Jr. 1978. Dynamics and succession of the phytoplankton in a tropical lake: Lake Lanao, Philippines. J. Ecol. 66:849–880.

Lewis, W. M., Jr. 1979. Zooplankton Community Analysis. Springer-Verlag, New York.

Likens, G. E., F. H. Bormann, R. S. Pierce, J. S. Eaton, and N. M. Johnson. 1977. Biogeochemistry of a Forested Ecosystem. Springer-Verlag, New York. 146 pp.

Lovejoy, T. E. 1975. Bird diversity and abundance in Amazon forest communities. The Living Bird 13:127–191.

Lovejoy, T. E. 1979a. Refugia, refuges, and minimum critical size: problems in the conservation of the neotropical herpetofauna. Pages 461–464 in W. E. Duellman, ed. The South American Herpetofauna: Its Origins, Evolution, and Dispersal. Monogr. Mus. Nat. Hist. Kans. No. 7. Lawrence, Kans.

Lovejoy, T. E. 1979b. Conservation beyond our borders. Nature Conservancy News 29(4):4–7.

Lovejoy, T. W., and J. M. Rankin. 1979. A checkered landscape: the implications of forest patch dynamics for forestry and reserve design. Paper presented at International Symposium on the Forest Sciences and Their Contribution to the Development of Tropical America, 11–17 October, 1979, San José, Costa Rica.

Lovejoy, T. E., and D. C. Oren. In press. Minimum critical size of ecosystems. In R. L. Burgiss and D. M. Sharpe, eds. Forest Island Dynamics in Man-Dominated Landscapes. Springer-Verlag, New York.

Lowe, R. G. 1975. Nigerian experience with natural regeneration in tropical moist forest. Background paper, Technical Conference on Tropical Moist Forest. Food and Agriculture Organization of the United Nations (FAO), Rome, Italy. 14 pp.

Lowe-McConnell, R. H. 1969. Speciation in tropical environments. Biol. J. Linn. Soc. 1:1–246.

Lowe-McConnell, R. H. 1975. Fish Communities in Tropical Freshwaters: Their Distribution, Ecology, and Evolution. Longman, New York.

Luther, J., and J. Rzóska. 1971. Project Aqua: A Source Book of Inland Waters Proposed for Conservation. Blackwell Scientific Publications, Oxford, England.

McNamara, R. S. 1979. Address to the Board of Governors by the President of the World Bank, Belgrade, Yugoslavia, October 2, 1979.

References

Marzola, D. L., and D. P. Bartholomew. 1979. Photosynthetic pathway and biomass energy production. Science 205:555–559.

Matheny, R. T., and D. L. Gurry. 1979. Ancient hydraulic techniques in the Chiapas highlands. Am. Sci. 67:441–449.

Matthew, K. M. 1978. Tropical botany: challenges, opportunities and responsibilities. Indian J. For. 1:1–8.

May, R. M. 1975a. Island biogeography and the design of wildlife preserves. Nature 254:177–178.

May, R. M. 1975b. The tropical rainforest. Nature 257:737–738.

May, R. M. 1979. Fluctuations in abundance of tropical insects. Nature 278:505–507.

Medina, E., R. Herrera, C. Jordan, and H. Klinge. 1977. The Amazon Project of the Venezuelan Institute for Scientific Research. Nat. Resour. XIII(3):4–6.

Medina, E., M. Sobrado, and R. Herrera. 1978. Significance of leaf orientation for leaf temperature in an Amazonian sclerophyll vegetation. Radiat. Environ. Biophys. 15:121–140.

Meijer, W. 1971. Regeneration of tropical lowland forest in Sabah, Malaysia, 40 years after logging. Malay. For. 23:204–229.

Meijer, W. 1973a. Endangered plant life. Biol. Conserv. 5:163–167.

Meijer, W. 1973b. Devastation and regeneration of lowland dipterocarp forests in Southeast Asia. BioScience 23:528–533.

Meijer, W. 1975. Indonesian Forests and Land Use Planning. University of Kentucky Bookstore, Lexington, Ky. 112 pp.

Melack, J. M. 1976. Primary productivity and fish yields in tropical lakes. Trans. Am. Fish. Soc. 105(5):575–580.

Meyer, J. L., and G. Likens. 1979. Transport and transformation of phosphorus in a forest stream ecosystem: a 13-year record. Ecology 60(6):1257–1271.

Miller, L. D. 1975. Butterfly conservation: the right way, the wrong way, the government way. Insect World Dig. 1975:2–9.

Molion, L. C. B. 1976. A climatonic study of the energy and moisture fluxes of the Amazon Basin with consideration of deforestation effects. Ph.D. dissertation, University of Wisconsin, Madison.

Montgomery, G. G., ed. 1978. The Ecology of Arboreal Folivores. Smithsonian Institution Press, Washington, D.C. 574 pp.

Mooney, H. A., O. Bjorkman, A. E. Hall, E. Medina, and P. B. Tomlinson. 1980. The study of the physiological ecology to tropical plants—current status and needs. BioScience 30(1):22–26.

Moreno Azorero, R., and B. Schvartzman. 1975. 268 plantas medicinales utilizadas para regular la fecundidad en algunos países de Sudamerica. Reproducción 2:163–183.

Mori, S. A., and L. A. M. Silva. 1979. The herbarium of the Centro de Pesquisas do Cacau at Itabuna, Brazil. Brittonia 31(2):177–196.

Myers, N. 1976. An expanded approach to the problem of disappearing species. Science 193:198–202.

Myers, N. 1979a. The Sinking Ark. Pergamon Press, Oxford, U.K. 307 pp.

Myers, N. 1979b. Islands of conservation. New Sci. 83:600–602.

Ng, F. S. P. 1978. Strategies of establishment in Malayan forest trees. Pages 129–162 in P. B. Tomlinson and M. H. Zimmermann, eds. Tropical Trees as Living Systems. Cambridge University Press, Cambridge, U.K.

Nicholls, Y. I. 1973. Source book: emergence of proposals for recompensing developing countries for maintaining environmental quality. Intern. Union Conserv. Nat., Environ. Policy and Law Paper No. 5. Morges, Switzerland.

Nielsen, U., and A. H. Aldred. 1976. Can tropical forest inventories benefit from recent developments in aerial photography? Pages 245–260 *in* Remote Sensing in Forestry. Proceedings of the Symposium held during the XVI IUFRO World Congress, Oslo, Norway. International Union of Forestry Research Organizations, Freiburg, Federal Republic of Germany.

NRC (National Research Council). 1970. Systematics in Support of Biological Research. National Academy of Sciences, Washington, D.C. 25 pp.

NRC (National Research Council). 1972. Soils of the Humid Tropics. National Academy of Sciences, Washington, D.C. 219 pp.

NRC (National Research Council). 1975. Underexploited Tropical Plants with Promising Economic Value. National Academy of Sciences, Washington, D.C. 189 pp.

NRC (National Research Council). 1977. Energy and Climate. National Academy of Sciences, Washington, D.C. 158 pp.

NRC (National Research Council). 1979. Tropical Legumes: Resources for the Future. Report of an ad hoc panel of the Advisory Committee on Technical Innovation, Board on Science and Technology for International Development, National Academy of Sciences, Washington, D.C. 231 pp.

NRC (National Research Council). 1980. Conversion of Tropical Moist Forests. National Academy of Sciences, Washington, D.C. 205 pp.

NSF (National Science Foundation). 1977. Long-Term Ecological Measurements. Report of a Conference, Woods Hole, Massachusetts, March 16–18, 1977. National Science Foundation, Washington, D.C. 26 pp.

NSF (National Science Foundation). 1978. A Pilot Program for Long-Term Observation and Study of Ecosystems in the United States. Report of a Second Conference on Long-Term Ecological Measurements, Woods Hole, Massachusetts, February 6–8, 1978. National Science Foundation, Washington, D.C. 44 pp.

Odum, H. T. 1970. Summary. An emerging view of the ecological system of El Verde. Pages I-191–I-281 *in* H. T. Odum and R. F. Pigeon, eds. A Tropical Rain Forest: A Study of Irradiation and Ecology at El Verde, Puerto Rico. Division of Technical Information, U.S. Atomic Energy Commission, Oak Ridge, Tenn.

Ojasti, J. 1968. Notes on the mating behavior of the capybara. J. Mammol. 49(3):534–535.

Ojasti, J. 1971. El chiguire. Asoc. Nac. Def. Natur. 1(3):3–10.

Ojasti, J., and G. Medina Padilla. 1972. The management of capybara in Venezuela. Pages 268–277 *in* Transactions of the 37th North American Wildlife and Natural Resources Conference. Wildlife Management Institute, Washington, D.C.

O'Neill, J. P., and G. R. Graves. 1977. A new genus and species of owl (Aves: Strigidae) from Peru. Auk 94(3):409.

Openshaw, K. 1974. Wood fuels the developing world. New Sci. 61:271–272.

Orozco-Segovia, A. D. L., and S. R. Gliessman. 1979. The *Marceno* in flood-prone regions of Tabasco, Mexico. Paper presented at the Symposium on Mexican Agroecosystems, Past and Present, organized for the XLIII International Congress of Americanists, Vancouver, Canada, 11–17 August 1979.

Payne, D. 1980. Trees and disease. New Sci. 85:12–14.

Pelzer, K. J. 1978. Swidden cultivation in Southeast Asia: historical, ecological and economic perspective. Pages 271–286 *in* P. Kunstadter, E. C. Chapman, and S. Sabhasri, eds. Farmers in the Forest. Economic Development and Marginal Agriculture in Northern Thailand. University Press of Hawaii, Honolulu. 402 pp.

Polhill, R. M. 1968. Tanzania. Acta Phytogeogr. Suec. 54:166–178.

Poore, D. 1976. The values of tropical moist forest ecosystems. Unasylva 28:112–113, 127–143.

References

Poore, D. 1978. Values of tropical moist forests. Pages 85–123 *in* Proceedings of Conference on Improved Utilization of Tropical Forests. Forest Products Laboratory, Forest Service, U.S. Department of Agriculture, Madison, Wis.

Population Reference Bureau. 1979. World Population Data Sheet. Washington, D.C.

Potter, G. L. 1975. Anthropogenic climate modification: Modeling the removal of tropical rain forests. Ph.D. dissertation, Lawrence Livermore Laboratory, University of California, Livermore.

Potter, G. L., H. W. Elsaesser, M. C. MacCracken, and F. M. Ruther. 1975. Possible impact of tropical deforestation. Nature 258:697–698.

Prance, G. T. 1975. Botanical training in Amazonia. Am. Inst. Biol. Sci. Educ. Rev. 4:1–4.

Prance, G. T. 1977. Floristic inventory of the tropics: Where do we stand? Ann. Mo. Bot. Gard. 64:659–684.

Prance, G. T., and T. S. Elias. 1977. Extinction is Forever. New York Botanical Garden, Bronx, New York. 437 pp.

Pye, V. I. 1976. New study: germplasm resources. ALS LifeLines 2(4):1–3.

Rappaport, R. A. 1968. Pigs for the Ancestors: Ritual in the Ecology of a New Guinea People. Yale University Press, New Haven. 311 pp.

Rappaport, R. A. 1978. Maladaptation in social systems. Pages 49–87 *in* J. Friedman and M. J. Rowlands, eds. The Evolution of Social Systems. University of Pittsburgh Press, Pittsburgh, Pa.

Rauh, W. 1979. Problems of biological conservation in Madagascar. Pages 405–421 *in* D. Bramwell, ed. Plants and Islands. Academic Press, New York.

Raven, P. H. 1975. The bases of angiosperm phylogeny: cytology. Ann. Mo. Bot. Gard. 62:724–764.

Raven, P. H. 1976. The destruction of the tropics. Frontiers 40(4):22–23.

Raven, P. H., and D. I. Axelrod. 1972. Plate tectonics and Australasian paleobiogeography. Science 176:1379–1386.

Raven, P. H., and D. I. Axelrod. 1974. Angiosperm biogeography and past continental movements. Ann. Mo. Bot. Gard. 61:539–673.

Raynal, J. 1979. Three examples of endangered nature in the Pacific Ocean. Pages 149–150 *in* I. Hedberg, ed. Systematic Botany, Plant Utilization and Biosphere Conservation. Almquist & Wiksell International, Stockholm.

Reichel-Dolmatoff, G. 1975. Cosmology as an ecological analysis: a view from the rain forest. Man 2(3):307–318.

Richard-Vindard, G., and R. Battistini, eds. 1972. Biogeography and Ecology of Madagascar. W. Junk, The Hague, the Netherlands. 765 pp. (Monographiae Biologicae, Vol. 21).

Richards, P. W. 1952. The Tropical Rain Forest. Cambridge University Press, London. 450 pp.

Richards, P. W. 1973. The tropical rain forest. Sci. Am. 229(6):58–67.

Richerson, P. J., C. Widner, and T. Kittel. 1977. The Limnology of Lake Titicaca (Peru–Bolivia), A Large, High Altitude Tropical Lake. Institute of Ecology Publication No. 14. University of California, Davis.

Robinson, M. H. 1978. Is tropical biology real? Trop. Ecol. 19(1):30–50.

Rollet, B. 1974. L'architecture des Forêts Denses Humides Sempervirentes de Plaine. Centre Technique Forestier Tropical, Nogent-sur-Marne, France. 298 pp.

Ruddle, G. 1974. The Yupka Cultivation System: A Study of Shifting Cultivation in Colombia and Venezuela. Ibero-Americana Vol. 52. University of California Press, Berkeley.

Ruttner, F. 1952. Planktonstudien der Deutschen limnologischen Sundaexpedition. Arch. Hydrobiol. Suppl. XXI:1–274.

Rzóska, J., ed. 1976. The Nile: Biology of an Ancient River. Monogr. Biol. 29. Dr. W. Junk bv, The Hague, the Netherlands. 417 pp.

Rzóska, J. 1978. On the Nature of Rivers with Case Stories of Nile, Zaire and Amazon. Dr. W. Junk bv, The Hague, the Netherlands. 67 pp.

Sailer, R. I. 1969. A taxonomist's view of environmental research and habitat manipulation. Pages 37–45 in Proceedings of the Tall Timbers Conference on Ecological Control by Habitat Management, No. 1. Tall Timbers Research Station, Tallahassee, Fla.

Salati, E., J. Marques, and L. Molion. 1978. Origem e distribução das chuvas na Amazônia. Interciencia 3:200–205.

Salati, E., A. Dall'Olio, E. Matsui, and J. R. Gat. 1979. Recycling of water in the Amazon Basin: an isotopic study. Water Resour. Res. 15(5):1250–1258.

Sarukhán, J. 1978. Studies on the demography of tropical trees. Pages 163–184 in P. B. Tomlinson and M. H. Zimmermann, eds. Tropical Trees as Living Systems. Cambridge University Press, Cambridge, U.K.

Sarukhán, J. 1979. Demographic problems in tropical systems. Pages 161–188 in O. T. Solbrig, ed. Plant Demography. Blackwell's, London.

Schneider, S. H. 1976. The Genesis Strategy. Climate and Global Survival. Plenum Press, New York. 419 pp.

Schnell, R. 1970. Introduction à la Phytogeographie des Pays Tropicaux. Vol. 1, Les Flores—Les Structures. Gauthier-Villars, Paris. 499 pp.

Simberloff, D. S., and L. G. Abele. 1976. Island biogeography theory and conservation practice. Science 191:285–286.

Simpson, B. B., and J. Haffer. 1978. Speciation patterns in the Amazonian forest biota. Ann. Rev. Ecol. Syst. 9:497–518.

Sioli, H. 1975. Tropical river: the Amazon. Pages 461–488 in B. A. Whitton, ed. Studies in Ecology. Vol. 2, River Ecology. University of California Press, Berkeley.

Smole, W. J. 1976. The Yanamamo Indians. University of Texas Press, Austin.

Soderstrom, T. R. 1979. The bamboozling *Thamnocalamus*. Garden 3(4):22–29.

Sommer, A. 1976. Attempt at an assessment of the world's tropical forests. Unasylva 28(112/113):5–27.

Soulé, M., and B. A. Wilcox. 1980. Conservation Biology: An Evolutionary-Ecological Perspective. Sinauer Associates, Sunderland, Mass.

Stanton, N. L. 1979. Patterns of species diversity in temperate and tropical litter mites. Ecology 60:295–304.

Stark, N., and C. F. Jordan. 1978. Nutrient retention by the root mat of an Amazonian rain forest. Ecology 59:434–437.

Stewart, P. 1976. Erosion in the weather machine. Commonw. For. Rev. 55(2):155–157.

Stumm, W., ed. 1977. Global Chemical Cycles and Their Alterations by Man. Dalhem Konferunzen, 61. Abakon Verlag, Berlin.

Swank, W. T., and J. E. Douglass. 1974. Stream flow greatly reduced by converting deciduous hardwood stands to pine. Science 185:857–859.

Talling, J. F. 1965. The photosynthetic activity of phytoplankton in East African lakes. Int. Rev. Gesamten Hydrobiol. 50:1–32.

Talling, J. F. 1966. The annual cycle of stratification and phytoplankton growth in Lake Victoria (East Africa). Int. Rev. Gesamten Hydrobiol. 51:545–621.

Talling, J. F. 1976. Water characteristics. Pages 385–402 in J. Rzóska, ed. The Nile: Biology of an Ancient River. Dr. W. Junk bv, The Hague, the Netherlands.

Tatum, L. A. 1971. The southern corn leaf blight epidemic. Science 171:1113–1116.

References

TIE (The Institute of Ecology). 1977. Experimental Ecological Reserves: A Proposed National Network. The Institute of Ecology, Indianapolis. 40 pp.

TIE (The Institute of Ecology). 1979. Long-Term Ecological Research: Concept Statement and Measurement Needs. Summary of a Workshop, Indianapolis, Indiana, June 25–27, 1979. The Institute of Ecology, Indianapolis. 27 pp.

Terborgh, J. 1974. Preservation of natural diversity: the problem of extinction prone species. BioScience 24:715–722.

Thompson, P. A. 1976. Factors involved in the selection of plant resources for conservation as seed in gene banks. Biol. Conserv. 10:159–167.

Thorington, R. W., Jr., and P. G. Heltne, eds. 1976. Neotropical Primates: Field Studies and Conservation. National Academy of Sciences, Washington, D.C. 133 pp.

Tinker, J. 1974. Why aren't foresters more conservationist? New Sci. 64(927):819–821.

Tomlinson, P. B. 1974. Breeding mechanisms in trees native to tropical Florida; a morphological assessment. J. Arnold Arbor. Harv. Univ. 55:269–290.

Trenbath, B. R. 1974. Biomass productivity of mixtures. Adv. Agron. 26:177–210.

UNESCO (United Nations Educational, Scientific, and Cultural Organization). 1974. Task Force on: Criteria and guidelines for the choice and establishment of biosphere reserves. Programme on Man and the Biosphere, MAB Rep. Ser. No. 22. Paris. 61 pp.

UNESCO (United Nations Educational, Scientific, and Cultural Organization). 1978. Tropical Forest Ecosystems. A state-of-knowledge report prepared by UNESCO/UNEP/FAO. Natural Resources Research XIV, UNESCO, Paris. 683 pp.

U.S. Interagency Task Force on Tropical Forests. 1980. The World's Tropical Forests: A U.S. Policy, Strategy and Program. U.S. Department of State, Washington, D.C.

Veprintsev, B. N., and N. Rott. 1979. Conserving genetic resources of animal species. Nature 280:633–634.

Villa Nova, N. A., E. Salati, and E. Matsui. 1976. Estimative de evaportranspiração na Bacia Amazônica. Acta Amazonica 6(2):215–228.

Wade, N. 1978. New vaccine may bring man and chimpanzee into tragic conflict. Science 200:1027–1030.

Went, F. W., and N. Stark. 1968a. Mycorrhiza. BioScience 18:1035–1039.

Went, F. W., and N. Stark. 1968b. The biological and mechanical role of soil fungi. Proc. Natl. Acad. Sci. U.S.A. 60:497–504.

Western, D., and W. Henry. 1979. Economics and conservation in third world national parks. BioScience 29(7):414–418.

Wetterberg, G. B., C. Soares de Castro, A. Tresinari, B. Quintao, and E. Rocha Porto. 1978. Estado atual dos parques nacionais e reservas equivalentes na america do sul—1978. Brasil Florestal No. 36, Min. Agric. IBDF, Brasilia. pp. 11–36.

Wetzel, R. M., R. E. Dubos, R. L. Martin, and P. Myers. 1975. *Catagonus,* an "extinct" peccary, alive in Paraguay. Science 189:378–381.

Wharton, G. W. 1964. First International Congress of Acarology keynote address. Acarologia 6:37–43.

Whitmore, T. C. 1975. Tropical Rain Forests of the Far East. Clarendon Press, Oxford, U.K.

Whitmore, T. C. 1976. Conservation Review of Tropical Rain Forests, General Considerations and Asia. International Union for the Conservation of Nature and Natural Resources, Morges, Switzerland. 116 pp.

Whitmore, T. C. 1978. Gaps in the forest canopy. Pages 639–655 *in* P. B. Tomlinson and M. H. Zimmermann, eds. Tropical Trees as Living Systems. Cambridge University Press, Cambridge, U.K.

Whitmore, T. C. 1980. The conservation of tropical rain forest. *In* M. Soulé and B. A. Wilcox, eds. Conservation Biology: An Evolutionary-Ecological Perspective. Sinauer Associates, Sunderland, Mass.

Wigley, T. M. L., P. D. Jones, and P. M. Kelly. 1980. Scenario for a warm, high CO_2 world. Nature 283:17–21.

Willis, E. O., and Y. Oniki. 1978. Birds and army ants. Ann. Rev. Ecol. Syst. 9:243–263.

Windsor, D. M. 1974 (Vol. II), 1975 (Vol. III), 1976 (Vol. IV). Environmental Monitoring and Baseline Data from the Isthmus of Panama. Smithsonian Institution, Washington, D.C. 252 pp.

Wolda, H. 1978. Some fluctuations in rainfall, food and abundance to tropical insects. J. Anim. Ecol. 47:369–381.

Woodwell, G. W., R. H. Whittaker, W. A. Reiners, G. E. Likens, C. C. Delwiche, and D. B. Botkin. 1978. The biota and the world carbon budget. Science 199:141–146.

World Bank. Forestry: Sector Policy Paper, February 1978. World Bank, Washington, D.C. 65 pp.

WMO (World Meteorological Organization). 1979. Proceedings of the World Climate Conference: A Conference of Experts on Climate and Mankind. World Meteorological Organization Report No. 537, Geneva, Switzerland.

Wyatt-Smith, J. 1963. Manual of Malayan Silviculture for Inland Forests. Malayan Forest Record No. 23, Kuala Lumpur, Malaysia. 400 pp.

Yanchinski, S. 1978. Brown planthopper stalks Vietnam's rice fields. New Sci. 80:342.

Yantko, J. A., and F. B. Golley. 1977. A World Census of Tropical Ecologists. Institute of Ecology, University of Georgia, Athens, Georgia, June 1977. 156 pp.

Yoneda, T., K. Yoda, and T. Kira. 1977. Accumulation and decomposition of big wood litter in Pasoh Forest, West Malaysia. Jap. J. Ecol. 27(1):53–60.

Zinke, P. J. 1977. Man's activities and their effect upon the limiting nutrients for primary productivity in marine and terrestrial ecosystems. Pages 89–98 *in* W. Stumm, ed. Global Chemical Cycles and Their Alterations by Man. Dahlem Konferenzen, Berlin.

Zucchi, A. 1975. La tecnología aborigen y el aprovechamiento agrícola de nuestras sabanas. Lineas (C. A. Luz Electrica de Venezuela y Compañias Afiliadas) 219.

Contributors

EDWARD S. AYENSU, Office of Biological Conservation, Smithsonian Institution, Washington, D.C.
HERBERT G. BAKER, University of California, Berkeley
ROBERT BLAKE, Natural Resources Defense Council, Washington, D.C.
ROBERT W. BRANDT, Forest Service, U.S. Department of Agriculture, Washington, D.C.
F. H. BORMANN, Yale School of Forestry and Environmental Sciences, New Haven, Connecticut
J. P. M. BRENAN, Royal Botanic Gardens, Kew, Richmond, Surrey, England
JOHN BROOKS, Division of Environmental Biology, National Science Foundation, Washington, D.C.
EBERHARD BRÜNIG, University of Hamburg, Federal Republic of Germany
JAMES T. CALLAHAN, Division of Environmental Biology, National Science Foundation, Washington, D.C.
WILLIAM CARLSON, Smithsonian Science Information Exchange, Washington, D.C.
THOMAS CROAT, Missouri Botanical Garden, St. Louis
DONALD R. DAVIS, Department of Entomology, National Museum of Natural Hsitory, Washington, D.C.
FRANCESCO DI CASTRI, Programme on Man and the Biosphere, United Nations Educational, Scientific, and Cultural Organization, Paris, France
PAUL EHRLICH, Stanford University, Palo Alto, California

JOHN EWEL, Department of Botany, University of Florida, Gainesville
CURTIS FREESE, Fish and Wildlife Service, U.S. Department of the Interior, Washington, D.C.
ALWYN GENTRY, Missouri Botanical Garden, St. Louis
LARRY E. GILBERT, Department of Zoology, University of Texas, Austin
VERNON C. GILBERT, National Park Service, U.S. Department of the Interior, Washington, D.C.
STEVE GLIESSMAN, Secretaria de Agricultura y Ganaderia, Colegio Superior de Agricultura Tropical, H. Cárdenas, Tabasco, Mexico
FRANK B. GOLLEY, Division of Environmental Biology, National Science Foundation, Washington, D.C.
ROBERT GOODLAND, Office of Environmental and Health Affairs, World Bank, Washington, D.C.
LESLIE D. GOTTLIEB, Department of Genetics, University of California, Davis
MALCOLM HADLEY, Division of Ecological Sciences, United Nations Educational, Scientific, and Cultural Organization, Paris, France
CALDWELL HAHN, World Wildlife Fund, Washington, D.C.
A. E. HALL, Department of Botany, University of California, Riverside
CHARLES HALL, Section of Ecology and Systematics, Cornell University, Ithaca, New York
GARY HARTSHORN, Tropical Science Center, San José, Costa Rica
O. HEDBERG, Institute of Systematic Botany, University of Uppsala, Sweden
LESLIE HOLDRIDGE, Tropical Science Center, San José, Costa Rica
HUGH ILTIS, Department of Botany, University of Wisconsin, Madison
M. JACOBS, Rijksherbarium, Leiden, the Netherlands
DAVID P. JANOS, Department of Biology, University of Miami, Coral Gables, Florida
TATUO KIRA, Department of Biology, Osaka University, Japan
SAM KUNKLE, Peace Corps, Washington, D.C.
J. P. LANLY, Forestry Department, United Nations Food and Agriculture Organization, Rome, Italy
G. E. LIKENS, Cornell University, Ithaca, New York
WILLIAM LONG, U.S. Department of State, Washington, D.C.
G. L. LUCAS, Royal Botanic Gardens, Kew, England, Richmond, Surrey, England
ARIEL LUGO, Institute of Tropical Forestry, Rio Piedras, Puerto Rico
BRUCE MACBRYDE, Office of Endangered Species, U.S. Fish and Wildlife Service, Department of the Interior, Washington, D.C.
ROGER MCMANUS, Endangered Species Scientific Authority, Washington, D.C.

Contributors

SAMUEL MCNAUGHTON, Biological Research Laboratories, Syracuse University, Syracuse, New York

WILLEM MEIJER, Department of Biological Sciences, University of Kentucky, Lexington

DIETER MUELLER-DOMBOIS, Department of Botany, University of Hawaii, Honolulu

BRAULIO OREJAS MIRANDA, General Secretariat, Organization of American States, Washington, D.C.

M. MATOS PEIXOTO, Conselho Nacional de Densenvolvimento Científico e Tecnológico, Brasília, Brazil

REIDAR PERSSON, National Board of Forestry, Jonkoping, Sweden

LAWRENCE PETTINGER, Applications Branch, EROS Data Center, Sioux Falls, South Dakota

PAUL CHAI PIANG-KONG, Forest Department, Kuching, Sarawak, East Malaysia

RIVALDO PINTO DE GUSMÃO, Superintendente de Recursos Naturais e Meio Ambiente, Rio de Janiero, Brazil

R. M. POLHILL, Royal Botanic Gardens, Kew, Richmond, Surrey, England

GHILLEAN PRANCE, New York Botanical Gardens, Bronx, New York

FRANK PRESS, Science and Technology Policy, Executive Office of the President, Washington, D.C.

STEPHEN B. PRESTON, School of Natural Resources, University of Michigan, Ann Arbor

ALBERT PRINTZ, U.S. Agency for International Development, Washington, D.C.

PAUL W. RICHARDS, Emeritus Professor of Botany, University of Wales, United Kingdom

BRUCE ROSS, Office of Technology Assessment, Congress of the United States, Washington, D.C.

IRA RUBINOFF, Smithsonian Tropical Research Institute, Balboa, Panama

ENEAS SALATI, Instituto Nacional de Pesquisas da Amazônia, Manaus, Amazonas, Brazil

JAY M. SAVAGE, Allan Hancock Foundation, University of Southern California, Los Angeles

RALPH H. SMUCKLER, Planning Office, Institute for Scientific and Technological Cooperation, Washington, D.C.

DONALD STONE, Department of Botany, Duke University, Durham, North Carolina

LEE TALBOT, World Wildlife Fund-International, Gland, Switzerland

DAVID B. THORUD, Forest Service, U.S. Department of Agriculture, Washington, D.C.

JOSEPH TOSI, Tropical Science Center, San José, Costa Rica

HAROLD E. WAHLGREN, Forest Service, U.S. Department of Agriculture, Washington, D.C.

LEONARD J. WEBB, Division of Plant Industry, Commonwealth Scientific and Industrial Research Organization, Indooroopilly, Queensland, Australia

TIMOTHY WHITMORE, Department of Forestry, University of Oxford, Commonwealth Forestry Institute, Oxford, England

GEORGE M. WOODWELL, Ecosystems Center, Marine Biological Laboratory, Woods Hole, Massachusetts